Earth Revealing—Earth Healing

Earth Revealing—Earth Healing

Ecology and Christian Theology

Edited by Denis Edwards

A Michael Glazier Book

THE LITURGICAL PRESS
Collegeville, Minnesota

www.litpress.org

A Michael Glazier Book published by The Liturgical Press

Cover design by Greg Becker.

Biblical quotations are from the New Revised Standard Version, copyright 1989, Division of Christian Education of the National Council of Churches in the United States of America.

1   2   3   4   5   6   7   8   9

Library of Congress Cataloging-in-Publication Data

Earth revealing, earth healing : ecology and Christian theology / edited by
    Denis Edwards.
        p. cm.
        "A Michael Glazier book."
        Includes bibliographical references and index.
        ISBN 0-8146-5951-9 (alk. paper)
        1. Human ecology—Religious aspects—Christianity. 2. Ecology—
    Religious aspects—Christianity. I. Edwards, Denis, 1943–

    BT695.5 .E273 2001
    261.8'362—dc21   00-064981

Contents

List of Illustrations

Acknowlegments

We thank the members of our editorial committee, Christine Burke, Jennie Teasdale, Duncan Reid, Lorna Hallahan, and Denis Edwards. Special thanks to Jennie Teasdale for keeping a record of our work, to James McEvoy for his careful reading of our texts, and to Paul Harrison for his willing work on the computer. We also wish to thank Louise Ritchard and Rev. Peter Tran Quong Tong for computer imaging.

We thank all those who participated in our initial forum on ecology and theology. Mark Twomey and Linda Maloney have been constant in their encouragement and support of this project, and we are very grateful to them and to all at The Liturgical Press.

A Meeting Place

In this picture of a meeting place,
people gather around campfires,
and they share food and culture.
I think our spiritual side is nurtured
by Mother Earth, because she brings
out the beauty in us. All Jesus put
on earth for us in beauty, and we
have to look after it—even the
colours of the flowers are there for us.

Hermy Munnich

Introduction

This book is the work of a diverse group of theologians committed to developing an ecological theology. The contributors to this volume share a conviction that the damage human beings are doing to the atmosphere, the seas, the rivers, the land, and the life-forms of the planet is extreme and deadly. We believe that it constitutes a crisis that demands all of humanity's wisdom, ingenuity, and commitment. The whole human community needs to be involved in the response to this crisis—young and old, women and men, farmers, politicians, gardeners, teachers, planners, scientists, engineers, artists, builders, cooks, and theologians.

In this context we are convinced that there is a responsibility upon us as theologians to rethink theology. In our view this is a small but significant part of the rethinking that the human community must do in its stance toward creation. We are well aware that some theological attitudes have contributed to exploitative attitudes and to heedless disregard for the good of the planet. We are also convinced that the biblical and theological tradition has resources that can be retrieved and developed as an ecological theology. It is our conviction that such a theology can contribute to the healing of our planet.

In 1997, in Adelaide, three theological colleges, representing the Anglican, Roman Catholic, and Uniting Churches, moved to a common campus. The three colleges had long been united in their academic program as the Adelaide College of Divinity and Flinders University's School of Theology. They were joined by a fourth college, Nungalinya College in Darwin. One of the fruits of the movement to a common campus was the impetus it gave to collegial research. Many of us involved in systematic theology decided to give priority to a common project in ecological theology. We were encouraged by the fact that our colleagues in biblical studies took a similar option and began work on the Earth Bible Project.

Together we have constituted a research center of Flinders University and the Adelaide College of Divinity called the Centre for Theology, Science and Culture. This center contributes to the dialogue between science and theology in cooperation with other research institutes, such as the Center for Theology and the Natural Sciences at Berkeley. It is also concerned with issues of spirituality and culture. Projects include not only the Ecotheology Project and the Earth Bible Project but also Exploring Australian Spirituality Through the Arts, Bringing the Archeological Sites of the Mediterranean to Adelaide, and Exploring the Spiritual Dimensions of Disability.

Those of us who agreed to work collaboratively on ecological theology decided, as a first step, to listen to a wider group of people who are committed to caring for the earth. We gathered gardeners, activists, ecofeminists, and scientists. Some, but not all, came from our various Church communities. They spoke of what they were doing and of what gave meaning to their work. We spent time reflecting on the experience that this group shared with us. Eventually we distilled it down to two main themes. First, they were saying that ecological commitment and their sense of God were already connected in an inner way, even if this is not often articulated. They spoke of the wonder and mystery of creation and of the Mystery that is both within and beyond it. We took this up as the theme of Earth Revealing. But they also spoke of their pain and anger at the damage done to the earth. They spoke of the need for undoing the damage, for a redemptive theology of the earth, and for a radically different ethos that would offer us an ecologically sustainable future. We took this up in the theme of Earth Healing.

In a recent book, *Women Healing Earth,* Rosemary Radford Ruether suggests that all of us need to be "less dogmatic and more creative about what is good and bad, usable and problematic, in our own cultural heritages." We need to discover what is liberating in our own traditions, avoiding the extremes of self-satisfied chauvinism on the one hand and the rejection of our own heritage as toxic waste on the other.[1] In this same book, Denise Ackermann and Tahira Joyner encapsulate the project of our own book when they write: "Earth-healing praxis is directed at restoring relationships between ourselves and all created life, and is infused by a spirituality which reverences the sacredness of all of creation."[2]

In this book we take up the theme of "Earth Revealing—Earth Healing" from the limited, partial position of people doing theology in an Australian context. This book makes no claim to be comprehensive. We see our contribution as a small part of a larger process that includes the work of other theologians and scholars in other disciplines, and as part of a

---

1. Rosemary Radford Ruether, ed., *Women Healing Earth: Third World Women on Ecology, Feminism and Religion* (Maryknoll, N.Y.: Orbis Books, 1996) 7.

2. Denise Ackermann and Tahira Joyner, "Earth Healing in South Africa: Challenges to Church and Mosque," in *Women Healing Earth,* 125.

worldwide collegial effort. This kind of collegiality is exemplified in another book from The Liturgical Press, from the faculty of the University of Portland, *All Creation Is Groaning,* edited by Carol J. Dempsey and Russell Butkus. Our offering is limited not only by our backgrounds but also by the particular research focus that each of us brings to the work. At the same time we see this book as forming a unity. The unity springs from the fact that we have spent many hours thinking through the issues together around the themes of Earth revealing and Earth healing. We represent different theological traditions and confessional stances, and we rejoice that our unity is a unity in difference. We have sought to welcome difference. Unity in difference is at the heart of the ecological theology we propose in this book, and it is clearly reflected in the way we have worked together.

It has been important to us that our way of working together should reflect some of the things that we are saying in the book. This process has involved an intense critical engagement with each other's work as well as much collegial support of one another. Each chapter was subjected to several rounds of peer review, collaborative discussion, and subsequent rewriting. This process of mutual refereeing was followed by a further careful review by our editorial committee and further reworking of each chapter. We are pleased to say that the process of mutual collaboration has resulted, not in uniformity of approach, but in each of our contributions becoming more distinctive.

The first chapter is Stephen Downs's reflection on landscape. He suggests that in responding to the ecological crisis, it is instructive to reflect on the relationship between culture and nature. More specifically, Downs argues that there are indissoluble bonds between nature and culture. Despite the fact that we habitually think and speak about ourselves as separate from the natural world, we are intimately linked with it. This is true not only biologically but also historically and culturally. It is true not only of indigenous cultures but also of Western culture, including Christian culture. In *Landscape and Memory,* Simon Schama digs below the surface of modern life and culture and finds many such cultural traditions, variously expressing the bonds between ourselves and nature. It is true that these connections have often taken the form of human beings exploiting or abusing nature. But it is also true that people have always acknowledged the sacredness of nature. This finding challenges the view that human beings are necessarily problematic in the ecosystem. In doing so, it provides some reason for hope. By re-membering and re-establishing our links with nature, we will be better able to respond to the crisis in nature that now confronts us.

In Chapter 2 Christine Burke writes on globalization and ecology, seeking signs of hope, positive movements in the global scenario and in the theological tradition. She argues that while the affluent world benefits from aspects of globalization, the negative impacts indicate the need for change in both attitudes and structures. Modernity has emphasized the individual and a rational, instrumental approach to our world. These priorities will have to change if we are to live creatively in the new millennium and pass on to our grandchildren the delights of a beautiful earth. An emerging consensus suggests that a renewed interest in spirituality and a valuing of mutuality, participation, and openness to difference and minority voices are key ingredients of an alternative culture. Burke surveys a number of key theological tenets to suggest how they can contribute to this new consensus. While such values are arguably central to the Christian endeavor, Christians and Church leadership must themselves embody them if they are to be credibly proclaimed by the Churches.

Denis Edwards's contribution is focused on the theology of the Holy Spirit as an ecological theology. His essay explores the traditional idea that the Spirit of God can be understood to be present to all creatures. It addresses the question: What, if anything, can be said theologically about the distinctive role of the Holy Spirit in ongoing creation? Edwards attempts an answer to this question that is both consistent with the Christian tradition and plausible in the light of contemporary evolutionary views. In a second step, he reflects on the claim of the theological tradition that the actions of the Trinity toward creation are undivided and one. This has often been taken to mean that there is no proper role for any one of the Trinitarian persons in creation. In response to this, Edwards argues that there is not only a distinctive but also a proper role of the Holy Spirit in creation. The chapter concludes with some reflections on the ecological consequences of this kind of theology of the Spirit.

Duncan Reid's chapter argues that although the term "incarnation" refers directly and literally to the Word's having become flesh, it has more often than not been understood to refer to the Word's having become human. Christology has jumped too hastily to consider the humanity of Christ, abstracted from the more immediate reference to flesh. In overlooking this reference, which links humanity to all other creaturely life on earth, Christology has been largely deprived of any ecological significance. Athanasius and Anselm are taken as two "core samples" of the tradition, exemplifying very different tendencies on this question. The article sets out to close the gap between the human and the flesh, to "enflesh" the human Christ, so as to retrieve a more earthy understanding of the incarnation. Two guiding principles are suggested for future work in Christology. Finally, a brief overview of a number of recent attempts to write an ecologically friendly Christology reveals these principles to be at work implicitly.

Patricia Fox addresses the ancient symbol of God as Trinity that is at the heart of Christian culture and tradition. She suggests that the Trinitarian picture of one-God-who-is-three shatters human categories, and she argues that recognizing and receiving God's otherness have an intrinsic connection with our capacity as Christians to recognize and receive non-human nature as the other. Drawing from the very different theologies of Elizabeth Johnson and John Zizioulas, Fox links the retrieval of God as Trinity with the restoration of non-human nature to its proper place within the sphere of the sacred. Both theologies are anchored in an understanding of the radical otherness of God and open up powerful and complementary visions of the Trinity as divine *koinonia* (communion). Together they provide a glimpse of how reclaiming the triune symbol of God can have profound existential consequences. Fox shows how key elements of their respective theologies contribute to an ethos and an ethic for Earth-healing.

Drawing on insights from disability theology, Lorna Hallahan explores our responses to the unloveliness that results from ecoinjustice. In "Embracing Unloveliness: Exploring Theology from the Dungheap," she suggests that we often discard the people, places, and creatures we deem incurable. They end up on the dungheap. Alternatively, building a theology from the dungheap allows us to start the work of healing. Theology from the dungheap does not propose solutions but speaks to our fears in a way that provides hope through effrontery. The effrontery is partisanship with unloveliness. This partisanship builds on an understanding of the metaphor of embrace as proposed by Miroslav Volf. Theology from the dungheap therefore seeks posterity for the weak through the enduring power of covenant relationships. Hallahan concludes by linking the work of ecoadvocacy with the four stages of the drama of embrace.

Anthony Lowes seeks to show how created persons contribute instrumentally to the final and enduring form of the universe. He offers a line of argument that complements the arguments used by most ecotheologians. He points to an anthropic principle that is the converse of anthropocentrism. Lowes traces an understanding of personhood found in recent writing on the Trinity, linking it with some significant postmodern, philosophical conceptualizations of the human person. He points to an alternative route to a greater valuing of non-personal creation. It is to be found in the very structure of personhood. Persons, human as well as divine, are essentially ecstatic. Their center of gravity is not within themselves but without. It is only by this decentering that persons have personal substance at all; otherwise they are merely conscious individuals. Lowes ascribes an instrumental function to created persons, divinized by grace. This function is the mechanism whereby created persons contribute in part to their own glorified bodiliness, that of other persons and, indeed, to the transfigured universe insofar as they have related to it in selfless love, in life and in death.

Lucy Larkin's chapter asks what it would take for any of us to create "right relationship" with nature, to learn to heal so that the flourishing of nature and ourselves can begin. The works of feminist theologians are utilized and pieced together as in a quilt. Larkin argues that healing occurs in the creative interaction between nature and ourselves. Domination in such relationships reduces the presence of the other and the creativity of connection, and prevents the healing that leads to flourishing. Ultimately our power to exercise non-dominative relationships with nature (human and non-human) comes from God.

Gregory Brett's reflection focuses on Christian hope (eschatology) and argues that it has a place at the table of ecotheologies. He suggests that true Christian eschatology is not stargazing or focusing on another, future world to the detriment of this world. Rather, a balanced eschatology has its sights firmly fixed in the present yet recognizes that we are in the middle of unfinished business. He briefly explores Karl Rahner's three guiding principles of eschatology and suggests that these principles can help us to engage in the conversation taking place on ecotheology. Armed with these principles, Brett engages in a brief conversation with three other participants in ecotheology, namely feminist, process, and sacramental theologies. Acknowledging their positive contribution, he offers some timely reminders from an eschatological perspective. These reminders both complement the work being done in these theologies and suggest further theological values that need to be preserved in order to sustain a balanced ecotheology.

In his chapter "Ecotheology as a Plea for Place," Phillip Tolliday attempts to situate the human within place rather than space. He notes the fundamental importance of place and the way in which it constitutes our selfhood. He recounts, in summary form, a brief history of place and suggests that the Judeo-Christian tradition displays a certain ambivalence about place that is not evidenced in the philosophy of Aristotle or in the stories that have come down to us through the dreaming from some Indigenous Australians. Despite a recovery of a sense of place in the twentieth century, a recovery that serves to reinforce the reader's sense of the importance of place, there remains the question of the ambivalence of our tradition toward place. This is an avenue to be further explored if we wish to embrace an integrated ecotheology.

James McEvoy's chapter begins with an examination of the roots of the ecological crisis in the thought and practices of the past three centuries. He turns to the analysis of modernity by Canadian philosopher Charles

Taylor and argues that the crisis has its roots in understandings of nature as neutral, existing to be made over by humans, and of human agency as disengaged, standing over nature in order to understand and master it. Taylor's analysis is then used to argue that an adequate solution to the crisis must be sought in renewed understandings of both human agency and nature that support the strengths and address the limitations of the modern worldview. In this context McEvoy argues that Karl Rahner's theology provides adequate understandings of human agency and nature with which to address our ecological troubles.

In Australia, as in most other developed countries, the Churches share in the community's struggle with a range of new questions of morality and public policy that are called "bioethical." Andrew Dutney reconsiders bioethics from the perspective of the ecological theology that has taken shape in our Project. He questions the common understanding of bioethics that places medical science at the center, orients the care of individuals around that center, cautiously admits limited consideration of human communities, and ignores entirely the natural environment. Dutney finds that this common understanding of bioethics is historically inaccurate and conceptually flawed, and has been largely overtaken by events at the end of the twentieth century. He also tracks the changing contribution of theology in the emergence of bioethics and finds that the approach represented in our Ecotheology Project is timely. It can help the Churches to participate constructively in the different kind of bioethics that present global circumstances demand.

Our artwork comes from Dunnilli Art, an enterprise workshop for Aboriginal and Torres Strait Islander people attached to Nungalinya College in Darwin. The Indigenous Australian artists whose works appear between the chapters are teaching and studying art at Dunnilli. Their art is drawn from personal experiences of life in their own lands. It gives expression to their sense of belonging to the land and of the connection between all things. Christine Burke and Jennie Teasdale were the point of contact between these artists and the ecological theology developed in this book. They spent time with the artists, listening to their stories, studying their art, and sharing reflections on this project. The contributing artists generously chose to share their work and stories with us and have given us permission to use them in this publication.

Contributors

Gregory Brett, C.M., teaches systematic theology at the Catholic Theological College, which is a member college of the Adelaide College of Divinity. His special interest is theological anthropology, particularly in the works of Karl Rahner and the philosophy of John Macmurray.

Christine E. Burke, I.B.V.M., teaches pastoral theology for Catholic Theological College in Adelaide and is director of the Church Ministry Program, which is responsible for the formation of non-ordained leadership within the Catholic Archdiocese of Adelaide. Her research interest is in public theology.

Stephen Downs is married with two children. He lectures in philosophy and theology and is dean of studies at Catholic Theological College, Adelaide. His research interests include contemporary philosophy and culture, world religions and theology and the arts.

Andrew Dutney teaches systematic theology in the Adelaide College of Divinity and at Flinders University. His recent research has focused on theological bioethics and he chairs the South Australian Council on Reproductive Technology. He is a minister of the Word in the Uniting Church in Australia.

Denis Edwards is priest of the Catholic Archdiocese of Adelaide and a lecturer in theology. In recent times his research focus has been on the theology of God and ecology, the interaction between science and theology and, currently, the Holy Spirit in creation.

Patricia A. Fox teaches at the Adelaide College of Divinity and Flinders University, and is presently President of the Institute of the Sisters of Mercy of Australia, based in Sydney. Her research interest is in the doctrine of God.

Lorna Hallahan is a social worker who is currently working toward a Ph.D. in theology at Flinders University, exploring an emancipatory theology of disability. She has worked as an advocate for and with people with disabilities.

Lucy Larkin, a religious education teacher from England, has taught theology, philosophy, and ethics to young adults. She submitted her Ph.D. thesis, which explores the theme of relationship for ecological, and particularly ecofeminist, theology.

Anthony Lowes is principal of Gleeson College. He is married, his wife also in leadership in Catholic Education. They have adult twin daughters. He is currently engaged in doctoral research at Flinders University on "The Trinity as Grammar of Reality."

James McEvoy is a Roman Catholic priest who works for Catholic Adult Education Service and writes a column for the *Southern Cross* magazine. His research interest is in the philosophy of Charles Taylor and the theology of Karl Rahner.

Duncan Reid is an Anglican priest who teaches systematic theology at the Adelaide College of Divinity and the Flinders University of South Australia. His research is in the doctrine of the triune God and the doctrine of creation. He is married and has three children.

Phillip W. Tolliday is an Anglican priest in the diocese of Adelaide. He has recently completed his Ph.D. thesis on "Providence By Way of Resurrection: A Hope Without Guarantee." He is currently preparing a project on the connection between radical theology and the holocaust.

The Flinders University and Adelaide College of Divinity Centre of Theology, Science and Culture.

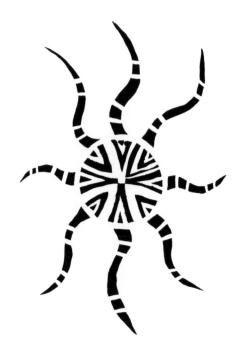

Mother Earth

The Earth is getting wrecked . . .
I don't like it.
More things are getting built.
More trees are getting chopped down.

You know
My people have always loved the earth . . .
From the start they loved it.
We can live without buildings,
Without air-conditioning and cars and that.
We communicate with animals,
With plants and wildlife . . .
We connect with them.
We love the bush and the fresh air . . .
You know, the main thing is the fresh air
And the sun.

Pollution can wreck it all for everyone
Trying to live on Mother Earth.

Yvette Talbot

Chapter 1

The Landscape Tradition

A Broader Vision for Ecotheology

Stephen Downs

Introduction

Human beings are part of the natural world. And yet we find it almost impossible to think or talk about nature without separating ourselves from it.[1] In *Landscape and Memory,* Simon Schama argues that it is to our mutual benefit to reflect on the indissoluble connections between ourselves and the natural world. As a historian, he believes that we should remember and celebrate the rich, ancient, and complex traditions that have embodied these connections. Only by doing so will we be able to respond effectively to the ecological crisis.[2]

In this chapter I will argue that it is instructive for theologians, and Christians in general, to reflect on Schama's position. It offers us a historical and cultural perspective on the ecological crisis and our response to it. While it confirms some of the strategies we have adopted, it challenges others. Like other writers in this volume, I have researched and written this chapter from an Australian viewpoint. This reflects my actual situation and may contribute some fresh insights to the global consideration of the ecological crisis. In the case of this chapter, there is an additional benefit. If valid, my arguments will confirm Schama's thesis—from a different perspective. His references are generally to Europe and North America.[3]

---

1. For an accessible discussion of this see George Seddon, *Landprints: Reflections on Places and Landscape* (Cambridge, U.K.: Cambridge University Press, 1997) 7–14.

2. Simon Schama, *Landscape and Memory* (London: Fontana Press, 1996) 13.

3. *Landscape and Memory,* and to a much lesser extent this article, refers to landscape painting traditions. For an interesting comparative study see Elizabeth Johns and others,

1

Shortly I will examine key aspects of Schama's thesis. As an introduction to his approach, I would first like to recall and reflect on a personal and concrete illustration of the intimate bond between human culture and nature.

Nature as Sacred

Recently my family celebrated Christmas at a "Family Mass" on the eve of the feast in our local parish. For some years now this liturgy has been held on spacious surrounding grounds outside the church. It had been a very hot day, and with several hours of daylight left, it was still uncomfortably warm in the sun. So we were pleased to arrange our fold-up chairs and picnic blankets in a shady grove nestled between the church and a fence on the boundary of the property. There were about a hundred families: parents and children, grandparents, uncles and aunts, and family friends. Some of the children came dressed as their favorite Nativity characters (many shepherds and several Marys). We gathered together, with room to stretch, around the altar table. There we praised and thanked God for coming among us in Jesus Christ.

As Roman Catholics, our liturgy took the form of a eucharistic celebration: God made present in the bread and wine. In view of our sylvan setting, it also seemed natural and appropriate for us to express in particular prayers our praise and thanks to the God of all creation for the sun, sky, and clouds; for the shade and cool gully breeze; for the earth beneath our feet and all the shrubs and flowers that it sustains; for our families, friends, and neighbors; for all the people rushing along the nearby road, preparing to celebrate Christmas in diverse ways.

Physically we are more accustomed to the dim lighting in our seventy-year-old church and the tight constraints of its pews. On this day, however, we were comfortable sitting randomly among a mix of native eucalyptus and imported conifers, the former affording us a "frail patterned sort of shade" and the latter, a genuinely cool "darkness."[4] Emotionally we usually feel constrained by the routine order of service. Among the trees, by contrast, we relaxed, as we would at a family picnic. I should add that we had the benefit of a sound system that enabled us to hear and join in with the presiding priest, the preacher, and the choir. This, however, was

---

New Worlds from Old: 19th Century Australian and American Landscapes (Canberra: National Gallery of Australia; Hartford, Conn.: Wadsworth Atheneum, 1998).

4. Some of the pine trees, native to Norfolk Island in the Pacific, were imported and planted over 150 years ago. That helps to explain why many Australians resist the notion of native-only gardens, etc. The contrasting description of gum trees and pine trees is taken from Murray Bail, *Eucalyptus* (Melbourne: Text Publishing, 1998) 14–15.

only one of many elements that converged to good effect. This Christmas Eve Eucharist was one of those occasions (for many of us, all too rare in liturgies) when everything seemed right.

Later I wondered how it was that we could feel so comfortable, religiously as well as physically, with this outdoor liturgy. It is not something that we experience very often. The few such liturgies that we do attend are usually held outdoors for practical reasons, and generally in less picturesque surroundings. Large school communities, for example, sometimes assemble for worship on a sporting field. The need to cater for larger than usual congregations at Christmas time is indeed one of the reasons why our local parish conducts its open-air services. But are there, I wondered, more profound factors at work here? It could be that the contemporary environmental movement has helped us to appreciate the general appropriateness of human activities in natural settings, though for most of us environmentalism is not overtly connected to religious ritual. Elements of our traditional liturgy do make some connections with the natural world. We offer to the "God of all creation" bread, "which earth has given," and wine, "fruit of the vine." As I have noted, however, we found it necessary to supplement these isolated references to nature with prayers of our own making. The much-needed retrieval of a doctrine of creation that balances the Christian doctrine of redemption is incomplete. It struggles with the strong tendency since the Reformation to focus on Christ's saving work *for us*. That creation (human and non-human alike) bears traces of Christ the creative Word of God is still inadequately reflected in most contemporary liturgies.[5]

In addition to all these factors, the reason why our relatively unusual experience of alfresco worship was so satisfying was because it resonated with one of the deeply embedded culture-nature traditions of which Simon Schama writes at length. As Schama retells the story, early Christians fled from the temptations of town or city life into the desert. Here they tested their faith against the elemental powers of darkness. But here, too, if their faith prevailed, they might be elevated to mystical union with God.

From at least the seventh century a European version of this practice (originally inspired by Hebraic experience) emerged: the retreat into the forest. Growing familiarity seems to have overcome the ancient Christian denunciation of heathen groves. The woodland hermitages and monasteries became so popular with pilgrims that those seeking solitude were forced into inaccessible places—in marshes or on mountaintops, for example.

---

5. Gerald O'Collins, "Jesus," *The Encyclopedia of Religion,* ed. Mircea Eliade (New York: Macmillan Publishing Company, 1987) 8:26. For examples of more expansive references to creation (nature) in ancient Christian liturgies, see R. N. McMichael, Jr., ed., *Creation and Liturgy* (Washington: The Pastoral Press, 1993).

Ultimately, however, it was the monastic gardeners who won the day. Even the densest and darkest woods, with "wildness, barbarism and paganism" could be turned into a "virtual paradise."[6] By the Middle Ages this developed into the tradition of the Christian monastery garden, "defined by its strong enclosing walls; the emblem of both Eden's prelapsarian self-sufficiency, and of the Virgin's immaculate conception: fertility without beasts or beastliness."[7]

That this tradition did not remain confined to the Christian monastic tradition is demonstrated in Schama's review of mid-nineteenth-century North American outdoor worship. It was commonplace for New Englanders to hold "tabernacle revival meetings" in open-air groves. And when the existence of the Big Trees of California became more widely known by European Americans, they were soon proclaimed "America's own natural temple." Routinely connected, by virtue of their antiquity, to Christ's birth, the Big Trees were part of what John Muir called the "Holy of Holies" in Yosemite.[8] Such views contributed significantly to the foundation of the first wave of formal environmental associations, such as Muir's Sierra Club. These also arose in Australia in the mid-nineteenth century and have at least historical links with the contemporary environmental movement.[9]

It is from this long and broad tradition of woods and gardens as places where Christians can properly encounter God that our Christmas Eucharist drew its full meaning. As it happens, our parish is attached to a monastery and retreat center. For over a hundred years, secluded from suburbia by tall fences and amidst trees and gardens, people from our city have left behind the concerns of daily life in order to focus on their relationship with God.[10] The founders of this monastery knowingly modeled it on a tradition virtually as old as the Gospel. Like our liturgy, the tradition of retreating to "natural surroundings" has a practical component: it is easier to leave the affairs of town or city life behind by physically retreating to the desert, woods, or secluded garden.

But as shown by Schama's examples, our relationship with nature is not simply practical. Over time Christians have recognized how appropriate and beneficial it is to undertake spiritual exercises in such places. Fur-

---

6. Schama, *Landscape and Memory,* 227.

7. Ibid., 534.

8. Ibid., 189f.

9. This is argued throughout Drew Hutton and Libby Connors, *A History of the Australian Environment Movement* (Cambridge, U.K.: Cambridge University Press, 1999).

10. My thanks to Brother Jeff Daly, C.P., archivist of the Passionist monastery, retreat house, and parish of Glen Osmond, South Australia.

ther, we have come to acknowledge the value of such places in themselves. According to Schama, "the cultural habits of humanity have always made room for the sacredness of nature."[11] That formal reflection on this aspect of Christian experience is comparatively underdeveloped needs to be noted and rectified by theologians. And as the illustrations show, the subjects of theological reflection need to include commonplace experiences of Christians, such as my Christmas Eve experience. With so much of our cultural heritage, its symbols and myths eroded by modern life and thought, it is in the "commonplace" that some of our deep traditions have managed to survive.[12]

History, Humanism, and Naturalism

Schama writes as a historian of culture and the environment, only addressing theoretical questions at any length in his introduction (less than 20 pages out of a total of 650!). Inasmuch as I concentrate on his fundamental arguments, referring only to the wealth of historical detail he provides to help illustrate my theological points, the reader should be aware that mine is not an entirely fair representation of Schama's cumulative argument.

For Schama, history consists in telling or retelling stories. In *Landscape and Memory* there are hundreds of stories. But in telling them he is making a point, ultimately a single point: we cannot separate the natural world from human culture. The truth of this derives from the indivisibility of nature and our perception of it: ". . . before it can ever be a repose for the senses, landscape is the work of the mind. Its scenery is built up as much from strata of memory as from layers of rock."[13]

Schama is not a philosophical idealist, believing that "to be is to be perceived." He does not question the independent existence of numerous ecosystems. There are few if any ecosystems, however, that remain totally unaffected by human existence. Whether through our use and abuse of them or our thinking and writing about them, all the earth's natural systems have in some way been altered by human existence.[14] Hence his consistent use of the term "landscape," originally referring to "a unit of human occupation."[15] George Seddon, an Australian multidisciplinary writer, takes much the same view: ". . . although they have a physical substrate, landscapes are also a cultural construct. The ways in which we read them,

---

11. Schama, *Landscape and Memory,* 18.
12. Compare Schama, *Landscape and Memory,* 14.
13. Ibid., 6–7.
14. Ibid., 7.
15. Ibid., 10.

perceive them, work them over, use them, evaluate them functionally, aesthetically, morally: these are all informed by our culture."[16]

Schama further argues that this intimate relationship between humanity and nature is not inherently bad. Acknowledging the sometimes terrible consequences for nature, he nevertheless believes our relationship with nature is something to celebrate.[17] In any event, without the conventions and constructions of culture, the natural world would be (for us) nothing but formless matter. Schama refers to the thought and work of the surrealist painter René Magritte: the world that we apprehend "needs a design before we can properly discern its form."[18] Culture provides that design.

Neurological case studies seem to confirm this view.[19] One such case, recorded by Oliver Sacks, is the basis of a recent Hollywood movie, *At First Sight*. Sacks's patient was a man who was born partially sighted but who lost even his limited sight in early childhood. As an adult he underwent an operation that, medically speaking, restored his sight. Despite this the man did not "see" the world as most of us do. For him "there was light, there was movement, there was colour, all mixed up, all meaningless, all a blur."[20] He had to learn how to see. Sacks concludes that we are not given the world; rather, "we make our world through incessant experience, categorization, memory, reconnection."[21] This seems to provide neurological support for the point made by Schama, Seddon, and Magritte about the active role played by culture in our very apprehension of nature.

A significant limitation of Schama's argument is that he rarely judges past human attitudes to nature and their consequences. Perhaps this is because he focuses on our present need to reassess the nature-culture relationship. Still he does refer to the "annexation of nature by culture," the need for "repair and redress," and the "ongoing degradation of the planet," thus acknowledging serious errors in the past conduct of the relationship.[22] But the principal purpose of his work is to demonstrate that there is more to the history of this relationship than the many "dismal tales" of recent environmental historians. Stories of "land taken, exploited, exhausted," of "reckless individualist[s]" and "capitalist aggressor[s]," of humans attack-

---

16. Seddon, *Landprints,* xv.

17. Schama, *Landscape and Memory,* 9–10.

18. Ibid., 12.

19. Details of some of these cases are included in a very rich account of "seeing" in Annie Dillard's deservedly famous meditation on living in the natural world, *Pilgrim at Tinker Creek* (New York: Harper's Magazine Press, 1974) ch. 4.

20. This quotation is from Sacks's original account, "To See and Not See," *An Anthropologist on Mars* (Sydney: Picador, 1995) 107.

21. Ibid., 108.

22. Schama, *Landscape and Memory,* 12, 14.

ing the land or treating it like a machine, of "ecological war"—these are not the only stories.[23]

A potential outcome of the work (and this is all Schama claims for it) is the recognition that we can learn from our history. He believes that we can find in our history cause for optimism in the face of the present ecological crisis. We can also find further motivation for urgent action, namely, rich, ancient, and complex traditions that will be lost unless we are able to solve these problems.[24] He strives to bring these traditions, often "hidden beneath layers of the commonplace," back into our conscious memories. Schama's own summary example is the "cathedral grove" of tourist cliché:

> Beneath the commonplace is a long, rich, and significant history of distinctive forms of Gothic architecture. The evolution from Nordic tree worship through the Christian iconography of the Tree of Life and the wooden cross to images like Caspar David Friedrich's explicit association between the evergreen fir and the architecture of resurrection may seem esoteric. But in fact it goes directly to the heart of one of our most powerful yearnings: the craving to find in nature a consolation for our mortality. It is why groves of trees with their annual promise of spring awakening, are thought to be a fitting décor for our earthly remains. So the mystery behind this commonplace turns out to be eloquent on the deepest relationships between natural form and human design.[25]

Such is the history that Schama uncovers and retells in his demonstration of the indissoluble links between humankind and the natural world. Careful study of our cultural heritage reveals that, far from involving the absolute repudiation of nature, it contains countless (and apparently universal) instances of the veneration of nature. Such study does not provide the concrete answers to our ecological problems or even strategies for addressing them politically. But it can help to motivate us and give us confidence by showing us that the history of our relationship with nature is one of intimate connection rather than inherent opposition, and has not been limited to the misuse and abuse of nature.[26]

One important characteristic of Schama's work is its relationship with the recent work of historians and others concerned with the environment. While *Landscape and Memory* contributes to the relatively new discipline of environmental history, it does not seek to displace human beings from the center of the earth's history. Schama's perspective is unashamedly

23. Ibid., 13.
24. Ibid., 14.
25. Ibid., 14–15. Each of the references here is treated in some detail in the main body of Schama's text.
26. Ibid., 17–19.

humanistic; his stories belong to "cultural history." He does not accept the mutual exclusivity of culture and nature, nor does he accept the endpoint of this approach—a critique of humanity, sometimes to the point of a generalized antipathy. This is not true of all environmental histories, some of which Schama compliments.[27] In any event, I think that Schama's history, with its emphasis on the first half of the culture-nature relationship, also complements them.

Schama's concerns with environmental history are more properly addressed, it seems to me, to those environmentalists who insist that the natural world can and must be considered from its own perspective. The Australian naturalist William Lines, for example, says of a jarrah forest: "It was always a jarrah forest, unstoried, beyond human intention and reach, and this will not be explained . . . . [C]ould I make sense of what we had seen, contrive unifying themes and achievements from a plethora of diverse and scattered observations without also imposing that story of purpose on nature? I had no answer."[28]

In the context of a discussion about attitudes to one of North America's natural wonders, Schama asks: "Would we rather that Yosemite, for all its overpopulation and overrepresentation, had never been identified, mapped and emparked?"[29] Presumably Lines and other "radical environmentalists" would answer "Yes!" This is problematic for a number of reasons. Sometimes research reveals more than first meets the eye. Schama refers to the case of Yosemite: ". . . the brilliant meadow-floor that suggested to its first eulogists a pristine Eden was in fact the result of regular fire-clearances by its Ahwahneechee Indian occupants."[30] Similarly, attempts by environmentalists in Australia to protect areas of "wilderness" have sometimes been questioned by the original Aboriginal inhabitants of those areas.

Schama's main difficulty with this type of environmentalism is with the *automatic* identification of human beings as ecologically problematic. While it is important to acknowledge that the "raw matter" of the natural world exists independently of the "landscapes" that we construct, it is very dangerous, for the human and non-human worlds alike and together, to

27. See Tom Griffiths' review of Stephen Pyne's *Vestal Fire: An Environmental History, Told Through Fire, of Europe and Europe's Encounter with the World* (Seattle: University of Washington Press, 1997) in *The Australian's Review of Books* (May 1998) 6–7. Schama compliments Pyne's work as well as that of William Cronon and Donald Worster in *Landscape and Memory,* 13.

28. William J. Lines, *A Long Walk in the Australian Bush* (Sydney: University of New South Wales Press, 1999) 82. This is otherwise a worthy and enjoyable book.

29. Schama, *Landscape and Memory,* 9.

30. Ibid.

idealize nature's wildness, its remoteness, its otherness. In addition to the negative account of human nature and culture that this romantic approach can engender, we human beings are left with little or no hope of creating solutions to our environmental problems.

Schama does, however have something in common with environmentalists of all types that is more than just a general concern for the state of the world: he shares their insistence on the need to reexamine, redescribe, and redefine the culture-nature relationship. The terms of their conclusions may differ. But some environmentalists at least will admit to sharing Schama's unease about the risks to social freedom posed by some suggested solutions to environmental problems, to being cautious about making a radical choice between redemption and extinction.[31] Schama would much prefer the position of environmental historian William Cronon: "The choice is not between two landscapes, one with and one without human influence; it is between two human ways of living, two ways of belonging to an ecosystem."[32] For this reason Schama's humanism needs to be understood in the context of his stated theme. It is his chosen perspective, not a conclusion about where value is to be found in the world. Using copious examples, he consistently acknowledges the value of both the human and other parts of the natural world.

This is so even when the stories he tells of people's relationship with nature contain a degree of ambiguity. Perhaps the most affronting of these tells of how the most barbaric regime in history was also one of the most ecologically conscientious:

> Arguably, no German government had ever taken the protection of the German forests more seriously than the Third Reich and its Reichforstminster Göring. Reluctant eleven-year-olds were turned into expert leaf-peepers through programs on forest ecology introduced into schools and were shown how the woodlands demonstrated the laws of biological competition and survival from the earwig to the eagle. Conservation was institutionalized through the creation of an entire administration run by the likes of Schönichen (who lectured on the subject at Berlin University from 1934 to 1936).[33]

Schama is quick to point out that this does not reveal any necessary link between environmentalism and totalitarianism. What it most clearly demonstrates is the tradition of an intimate connection between culture and nature. Nazi environmentalism was carefully constructed on a history of German devotion to the forest that Schama documents as far back as the

---

31. Ibid., 18, 119.
32. Quoted in Seddon, *Landprints*, 23.
33. For this quotation and the following see Schama, *Landscape and Memory*, 119.

wars between Rome and the German tribes around the time of Christ's birth. The form that this connection took in Nazi Germany—between German myths of the forest and militant nationalism—is unpleasant in the extreme. It nonetheless confirms the existence of ancient and continuous traditions connecting us with nature.

A less painful but equally ambiguous example of human regard for the sacredness of nature is found in the history of the sacred mountain *(sacro monte)*. Even when restricting the story to lifelike reproductions of Calvary, Schama is able to go back to the thirteenth century. At Monte Verna in northern Italy, St. Francis of Assisi was said to have received a revelation that it was these very rocks that were rent asunder at the moment of Christ's death.[34] Thus with honest religious enthusiasm and naïve literalism began a tradition of sacred mountains that spread across Europe for several succeeding centuries. In the seventeenth century Mont Valérien, just west of Paris, attracted enormous numbers of people, including Louis XIII and Anne of Austria. A century later, with at least five chapels and nine "stations," it was still able to attract a fervent (if inconsistent) Jean-Jacques Rousseau.[35] The most recent and perhaps the last example is "Holy Land USA." In the 1950s a thirty-two-foot cross made of stainless steel and lit with neon was erected on a hill overlooking Waterbury, Connecticut: "In 1958 HOLY LAND USA, announced by giant capital letters—the beatific rebuttal of the HOLLYWOOD sign—opened for business. At its peak . . . [a]bout two thousand visitors a day wandered around the hundred odd quarter-size buildings representing Bethlehem, Jerusalem and Nazareth."[36]

However mixed the motives of the developers of these attractions, the fact remains that they tapped into age-old "Judaic, Christian, and Muslim traditions . . . full of mountain epiphanies and transfigurations—on Horeb, Aarat, Moriah, Sinai, Pisgah, Gilboa, Gibeon, Tabor, Carmel, Calvary, Golgotha, Zion."[37] In inviting us to remember such traditions, along with all their ambiguous associations, Schama's work does not mourn the loss of the union of nature and culture. Instead, with "eyes wide open," it calls us to acknowledge their indissoluble bonds, and in so doing remain faithful to both.[38]

34. Ibid., 436.
35. Ibid., 440f.
36. Ibid., 446.
37. Ibid., 410.
38. Compare Schama, *Landscape and Memory*, 19.

Anthropocentric or Biocentric?

Because of its unmistakable humanism, Schama's position is likely to be regarded as anthropocentric. This reading of Schama is reinforced by his eagerness to denounce the misanthropy of some environmentalists. If my understanding of Schama's humanism is correct, however, this would be a misreading of *Landscape and Memory*. In repeatedly and consistently detailing examples of people attributing value to the natural world and finding value in it (it is nearly impossible to distinguish these actions in practice), Schama's work promotes a non-anthropocentric outlook. This is certainly true if "anthropocentrism" is understood as the belief that all values reside in or are derived from the human.[39] One of the key questions in environmental philosophy today is whether or not we can affirm the existence of intrinsic values in nature (that is, outside human nature). Through a myriad of examples Schama does precisely this. I do not see how his repeated illustrations of human belief in the sacredness of nature can be read otherwise.

Schama's humanistic outlook does rule out the set of positions known as deep ecology and biocentric egalitarianism. But it is entirely consonant with the biocentric approach of Charles Birch and an increasing number of environmental thinkers, according to which all organisms have intrinsic value but not all organisms have equal intrinsic value. Birch's particular account has the added advantage of being presented in a more conciliatory and generally positive manner.[40] Against some more radical non-anthropocentric positions, he argues that human judgments, such as the differentiation of grades of intrinsic value, are *necessarily* made from the human point of view. Accordingly, we need to distinguish between the value of the non-animate and the animate world and, within the latter category, between grades of intrinsic value.[41] Far from precluding us from adopting a sympathetic attitude toward the non-human world, this position requires us to acknowledge the rights of all living beings. But it also insists that unless we are willing to make such distinctions, our non-anthropocentric ethic will be confused and impractical.

---

39. *The New Shorter Oxford Dictionary* (1993) defines "anthropocentrism" more broadly, regarding humanity as the central fact of the universe. Schama might agree with this.

40. For this and what follows see Charles Birch, *Regaining Compassion for Humanity and Nature* (Kensington, N.S.W.: University of New South Wales Press, 1993) 99–106.

41. Unlike James McEvoy and some other writers in this volume, Birch is happy to use the notion of instrumental value in connection with the non-animate world: "Its value is instrumental to us and to other living creatures" (Birch, *Regaining Compassion,* 105). I suspect that McEvoy's argument is aimed more at those who contrast the intrinsic value of the human sphere with the merely instrumental value of the entire non-human sphere. Birch clearly attributes intrinsic value to non-human life.

Perhaps most importantly of all, the biocentric approach of Birch and others also leads us to accept human responsibility for the fate of our world. Christian ecotheology can, I think, learn from Schama and Birch here. It too must address the historical (and sometimes hysterical) charge of anthropocentrism. According to a typical allegation, "Christian arrogance toward nature" (to quote the oft-quoted Lynn White) is the major source of the contemporary ecological crisis.[42] Even those who acknowledge much greater complexity in Christian tradition generally concede that the most dominant Christian traditions have been anthropocentric, in the strong sense of the term. John Passmore characterizes these as "Man [*sic*] as Despot" and "Stewardship and Cooperation with Nature."[43]

The first tradition is now generally attributed to a Christian misreading of Genesis 1:26-28 and 9:2-3; the second, to a reading of Genesis 2:15. A third tradition is also sometimes acknowledged. Most famously found in St. Francis of Assisi, it maintains that "we are fellow companions of other creatures, all of whom rejoice in the beneficence of God."[44] It can be supported by reference to Genesis 1 as a whole (read with its original theocentric intent) and several of the psalms, where all the earth "sings a new song." One of the major theological responses to the ecological crisis, particularly to allegations of Christian culpability, has been to revisit one or more of these traditions. A significant number of theologians are trying to elevate the third tradition, suitably developed, to a more prominent position in Church and society.

Schama's work questions the adequacy of even the revised account of traditional Christian views of nature. Certainly Christian theology has been strongly anthropocentric in character. Its views have been widely influential—for example, in the emergence of modern science and technology. In this way Christian beliefs and attitudes are implicated in the ecological situation. Schama's historical excavations, however, have unearthed numerous instances over two millennia of human customs, practices, and conventions that demonstrate the existence of a strong parallel tradition that is biocentric in character.

How is it possible to reconcile these discoveries with the view held by Passmore and others that for Christianity nature is not sacred? According to this position, Christians believe that God alone (and that which is specifi-

---

42. Lynn White, Jr., "The Historical Roots of Our Ecological Crisis," *Science* (March 1967) 1204.

43. John Passmore, *Man's Responsibility for Nature* (London: Duckworth, 1974). These are the titles of chapters 1 and 2.

44. Birch, *Regaining Compassion*, 92.

cally dedicated to God) is sacred.[45] This is a very narrow use of the term "sacred." If it means that Christians do not think it appropriate to *worship* nature, then it is true. Few if any of the Christians in Schama's stories worshiped trees, rivers, or mountains, but all of them saw it as right and fitting to venerate and respect them, ultimately because of the natural world's connection with God. It seems that they had a better grasp than contemporary philosophers and some theologians of what Mircea Eliade describes as the indissoluble connection between the supernatural and the natural.[46]

Landscape and Memory reminds us that culture is an extremely complex phenomenon. Theoretical categories, including the philosophical, theological, and scientific, are insufficient to comprehend its richness. In the specific case of religion, for example, while the doctrinal and philosophical dimensions are important, they do not capture the fullness of a living tradition. There are other important dimensions. Ninian Smart's account of religion lists these as the practical and ritual, the experiential and emotional, the narrative or mythic, the ethical and legal, the social and institutional, and the material dimensions.[47] Examples of most of these can be found in *Landscape and Memory*. Very few can be found in philosophies or theologies of nature. They are almost never given the prominence of the views of experts. John Passmore, for example, does refer to "popular attitudes to animals and to the life around them," noting that these "were by no means necessarily the same as those embraced by philosophers and theologians."[48] But this interesting point is relegated to an early footnote and is never referred to again.

I am not suggesting that theologians abandon the examination of key Christian doctrines in their efforts to develop a better theology of nature. Indeed, the critical reexamination of doctrines, reflecting our greater historical and hermeneutical awareness, is supported by my argument. A good example of this is the reconsideration of what has been called the "sacramental principle," the traditional Catholic belief that everything is capable of manifesting and communicating the divine. Writing about the emergence of ecofeminist theology, Susan Ross concludes: "A sacramental view of the cosmos—that is, a vision that sees all of creation and

---

45. Passmore, *Man's Responsibility for Nature,* 10–11, dismisses a more generous assessment of the biblical attitude to nature in an early work by Charles Birch, preferring his own exegesis of Genesis and the support of theologian Charles Davis.

46. Mircea Eliade, *The Sacred and the Profane: The Nature of Religion* (New York: Harcourt, Brace & World, Inc., 1959) 117–118.

47. Ninian Smart, *The World's Religions: Traditions and Modern Transformations* (Cambridge, U.K.: Cambridge University Press, 1995) 10–21.

48. Passmore, *Man's Responsibility for Nature,* 10.

non-human life as sacred—is the natural and logical outcome of a sacra-mental faith."[49]

Several of the essays in this volume also take this general approach, generating interesting and potentially useful findings.[50] There is, however, a tendency among theologians (as was noted about philosophers above) to limit their reflections on, and reconsideration of, the Christian tradition to doctrinal issues. I am suggesting that Christian doctrine is not the only ap-propriate subject for theological reflection and retrieval. My reading of *Landscape and Memory* suggests that popular traditions and practices, not just those sanctioned by the "experts," also merit further attention.

Whose Culture? Which Traditions?

Perhaps the clearest implication of Schama's work is articulated in this question: Whose culture and which traditions should we explore for an understanding of the relationship between culture and nature, an under-standing that can inform an effective response to the ecological crisis? Schama argues against those environmentalists who think Western civi-lization (for some, all civilization) not only caused our present predica-ment but that it is also powerless to save us from it. They argue that we need to find new "creation myths."[51] The most commonly proposed source for these new myths is other native cultures. A popular example of this ap-proach is David Suzuki and Peter Knudson's *Wisdom of the Elders,* which details nature myths of the indigenous peoples of North America.[52]

Schama's concern is not with the adequacy or validity of particular cul-tures and traditions; rather, he rejects the claim that often accompanies such works that we need to turn to someone else's culture and traditions for so-lutions to our ecological problems. Suzuki, for example, boldly claims that every immigrant to North America since Columbus has viewed the land purely as a commodity; that is the reason why we need a radically different way of relating ourselves to the land.[53] *Landscape and Memory* provides a

---

49. Susan A. Ross, *Extravagant Affections: A Feminist Sacramental Theology* (New York: Continuum, 1998) 178. Her account of the "sacramental principle" begins on page 34. A handy compendium of articles on the religious aspects of ecofeminism is Mary Heather MacKinnon and Moni MacIntyre, eds., *Readings in Ecology and Feminist Theology* (Kansas City: Sheed & Ward, 1995).

50. The essays by Gregory Brett (Chapter 9), Denis Edwards (Chapter 3), Patricia Fox (Chapter 5), Anthony Lowes (Chapter 7), and Duncan Reid (Chapter 4) are all of this type.

51. Schama, *Landscape and Memory,* 13.

52. David Suzuki and Peter Knudson, *Wisdom of the Elders: Sacred Native Stories of Nature* (New York: Bantam Books, 1993). The original subtitle was *Honouring Sacred Na-tive Visions of Nature.*

53. Ibid., xxvii and xliv.

host of counter-examples to this claim, a tiny fraction of which I have summarized in this article. It demonstrates the truth of Schama's thesis that Western culture has more than enough of its own native traditions to help it respond to the current emergency. One such tradition that has been rediscovered by many Christians in recent years is Celtic Christianity.[54]

Looking to our own traditions, whatever they may be, is not to deny the value of other traditions. It is simply to focus on that with which we are familiar. In Australia, for example, the turn to other indigenous cultures has led many to look to Australian Aboriginal cultural traditions. One attraction is that in traditional Aboriginal life, it seems, the distinction between nature and culture was almost completely dissolved.[55] Attractive as that may be to non-Aboriginal Australians in the present climate, it is questionable whether we can even understand this way of being in the world. As one writer put it: "It is hard to imagine . . . a religious system that is further removed from Judaeo-Christianity."[56]

There are, in addition to Schama's concerns, serious moral problems with the turn to other cultures, especially when it involves the appropriation of traditions from an oppressed culture.[57] The first is the possibility of outright theft: making up for a (perceived) deficiency in one's own culture by taking from another. In Australia this is particularly pernicious because it comes on top of, and in spite of our growing recognition of, the theft of Aboriginal land, art, and children.[58] The appropriation, by various means, of Aboriginal land and art at least has a strong component of financial profit.

It is also possible to exploit the spiritual traditions of a people. This can even be the effect of good intentions: for example, attributing Aborigines with a special responsibility for the land (as sacred). While this might seem to honor Aborigines' connection with the land, it can also marginalize them

54. See, for example, Ian Bradley, *The Celtic Way* (London: Darton, Longman and Todd, 1993), especially chapter 3, "The Goodness of Nature."

55. William J. Lines, *An All Consuming Passion: Origins, Modernity and the Australian Life of Georgiana Molloy* (St. Leonards, N.S.W.: Allen & Unwin, 1993) 11–12.

56. Max Charlesworth, *Religious Inventions: Four Essays* (Cambridge, U.K.: Cambridge University Press, 1997) 77.

57. I am indebted to the concise summary of these issues in Roslynn D. Haynes, *Seeking the Centre: The Australian Desert in Literature, Art and Film* (Cambridge, U.K.: Cambridge University Press, 1998) 279–280. A fine study in its own right, this book is a valuable Australian addendum to *Landscape and Memory*. Where Schama explores traditions connected with forests, rivers, and mountains over two millennia in Europe and North America, Haynes explores desert traditions from two hundred years or so of European settlement in Australia.

58. The term "stolen children/generation" refers to the practice of white authorities, including Christian Churches, of putting Aboriginal children of "mixed descent" into institutions. It was part of an official policy of assimilation.

by "relegating" them to the spiritual sphere, and so deprive them of an equal place in the socioeconomic and other material spheres. Furthermore, this approach can effectively relieve non-Aboriginal people of their responsibility to care for the land. In other words, it can disenfranchise Aboriginal people materially and disenfranchise non-Aboriginal people spiritually.[59]

Clearly there are grave intellectual and moral dangers in ill-considered appropriations of other indigenous cultural traditions. Still, given Christian theology's significant commitment to dialogue with other religious traditions in general, it would be perverse for theologians to ignore the traditions of the indigenous people of their own land. Schama's few references to non-European nature traditions are usually presented as contrasting approaches. For example, he compares the monumental conquest of Mount Rushmore by Europeans, in which landscape became manscape, with the indigenous Lakota people's desire to honor the Great Spirit, Wakonda, who is "indistinguishably embedded with the rock and the scree."[60]

A similar contrast exists in different attitudes toward one of Australia's most famous landforms, the monolith now known as Uluru (formerly Ayers Rock). The government transferred legal ownership of this sacred rock to its traditional custodians in 1985. Delighted to share the vision of Uluru with anyone prepared to journey to the central Australian desert out of which it rises some 348 meters (1140 feet), the local Aboriginal people nevertheless ask that people refrain from climbing it. Its formidable height, however, is matched by the determination of many tourists to do precisely that. Still, it remains true, as Schama claims, that knowledge of other cultures can help us to appreciate the limitations (and sometimes the strengths) of our own traditions.

Such knowledge may even help us to understand the origin and nature of our own traditions. Schama discusses at some length, for example, the realization by European anthropologists that elements of pagan animism survived in Judeo-Christian thought and practice. While some saw this as discrediting Christian belief, others saw it as lending greater force to religious belief. Agreeing with the latter, Schama believes that myths from different cultures can be "highly complex systems of understanding, with power to generate and determine social behavior, rather than the other way about."[61]

The point I wish to emphasize here is that Christians of earlier eras did engage with other cultures and traditions, and Christians today continue to

59. Haynes, *Seeking the Centre*, 279–280. See also David Tacey, *Edge of the Sacred: Transformation in Australia* (North Blackburn, Vic.: HarperCollins, 1995) 8–10.

60. Schama, *Landscape and Memory*, 398–399.

61. Ibid., 209.

benefit from that engagement. Our understanding of the Cross, for example, in all its richness and complexity, is inconceivable without this connection.[62] That is why some philosophers and theologians think it not only possible but imperative for us to engage in dialogue with other cultures and traditions. This must surely include the indigenous cultures of one's own land. In view of the continuing history of oppression, great care must be taken to ensure that this does not become another expression of colonialism.

What forms might such dialogue take? What might be its outcomes? While it, too, must address some complex problems, the art world provides us with some clues. For example, it is clearly inappropriate for non-Aboriginal artists to adopt the distinctive style of Aboriginal painters, such as the dot and circle designs of the Papunya school. Only Aboriginal people can create "Aboriginal art." The same must be said about "Aboriginal theology." But non-Aboriginal people can learn from Aboriginal culture such things as the possibility and value of a spiritual relationship with the land. Roslynn Haynes, among others, has identified a number of non-indigenous contemporary desert artists whose work is informed by this insight.[63] She quotes, for example, the English-born artist John Wolseley:

> My work is an involved form of contemplation. If you live in nature for a time, you have almost mystical experiences. You get into a state of lyrical excitement and become part of the things that are happening around you: the grass quivering, the birds singing. On a sunny day there is an extraordinary feeling of energy, as light dazzles, bees buzz, birds dart, lizards slither. Nature moves through your veins in a spiritual way, and then the work of art flows out like birdsong.[64]

Wolseley has created an artistic vision and style out of his own experience. In his case that includes some knowledge of the Christian Desert Fathers and Zen Buddhism, in addition to Aboriginal culture and art. His paintings do not resemble the works of Aboriginal artists. But Wolseley acknowledges the influence of Aboriginal culture, and through his work he and the viewers of his paintings can converse with it, specifically about the interaction between people and the land.[65] That is the kind of dialogue that Christian theology should have with Aboriginal culture.

---

62. Ibid., 214–226, for Schama's treatment of "the verdant cross."

63. Haynes, *Seeking the Centre,* ch. 13, discusses the work of John Coburn, John Olsen, John Wolseley, and Mandy Martin. For the point I am making here see page 288.

64. Haynes, *Seeking the Centre,* 281.

65. Ibid., 257. A color reproduction of one of Wolseley's paintings can be found in Haynes, following page 208. For a more detailed review of this painter's work see Sasha Grishin, *John Wolseley: Land Marks* (North Ryde, N.S.W.: Craftsman House, 1998).

Landscape and Memory shows us that rethinking our relationship with nature involves rediscovering lost and hidden traditions of our own. But as noted earlier, the intimate connection between culture and nature is a widespread and perhaps universal belief. So conversing with other cultures may well reinforce and complement our own native traditions.[66] My reading of Schama's work further suggests that when looking for points of contact between Christian and other cultural traditions, we must look beyond biblical texts and doctrinal formulations.

A Broader Vision for Ecotheology

Theology contributes to the present global discussion of the nature-culture relationship and how best to respond to the ecological crisis by ensuring that the conversation includes the vital religious dimension of human culture. That is one of its roles as an intellectual discipline. It is also part of a Gospel imperative. In the words of Paul Tillich: "to show to the people outside the Church that the symbols in which the life of the Church expresses itself are answers to the questions implied in their very existence as human beings."[67] Tillich wrote this in appreciation of the valuable insights of humanistic existentialism. Those insights remain valid.

We are now much more aware, however, of the need to extend our understanding of human existence to include the whole of the natural world and of our relationship with it. My reflections on Simon Schama's *Landscape and Memory* have led me to reconsider from the perspective of my own cultural traditions the sacredness of nature; to develop a humanism that is able to value the non-human world for itself; and to be more searching and sensitive in my investigation of my own and other cultural traditions. Applying these reflections to ecotheology will result in a broader vision of ecotheology—its scope, task, and methods. The contribution of ecotheology to our understanding of ourselves and our world and how we should respond to our present ecological difficulties can be even more fruitful.

---

66. Rosemary Radford Ruether draws a similar conclusion in a reflection on ecofeminist writings from the Third World. She also believes that all of us, in the North and the South, need to acknowledge the complexity of our traditions about culture and nature. See R. R. Ruether, "Introduction," *Women Healing Earth: Third World Women on Ecology, Feminism, and Religion*, ed. R. R. Ruether (Maryknoll, N.Y.: Orbis Books, 1996) 1–8.

67. Paul Tillich, *Theology of Culture,* ed. R. C. Kimball (London: Oxford University Press, 1975) 49.

The Land

This shows how most Aborigines live.
Without the Land they have nothing.
They love the Land.

This shows how my ancestors lived.
See, there's people with spears
Looking for food.
And there's bush weed in the billabong
Where they go hunting for file snakes.

Look at the hands of my ancestors.
Find them on cave walls.
Shows how old you are
By the size of your hand.

You know
Everybody in the whole world
Needs to have more respect for our Land
And for the whole earth.

Tami Munnich

Chapter 2

Globalization and Ecology

Christine E. Burke, I.B.V.M.

I sit at my desk to write this article, disturbed by the impact of globalization on the economies and ecologies of the world. And then I reflect. My computer was made in Japan. I e-mail a friend in Belgium with a plea for helpful references. I search a website in the United States for articles. The clothes I wear were made in China. Inescapably, I benefit from globalization. I write as a member of "our" affluent society and a member of "our" Christian community.

This volume develops a positive Christian engagement with ecology. While the authors recognize that Christianity shares some historical responsibility for dismissive attitudes toward nature, we are choosing to highlight ways in which the ecological crisis challenges those within the Christian tradition to contribute to healing the earth and its inhabitants. Some might ask, What has globalization to do with the earth healing or revealing, let alone with theology? The terms "ecology" and "economy" share a Greek root, *oikos,* or "household." Eco*logy* is about understanding our planetary and cosmic household, which is the home we cannot exit. Eco*nomics* points to the laws and organization of that household, our stewardship of its resources. "Ecumenism" shares the same root, capturing the sense of the whole inhabited earth as one household of God; so this ecumenical volume addresses the economy as part of the ecological crisis. A dialogue between globalization and the Churches challenges citizens who belong to the Christian community to uncover directions we can pursue in hope.

Amartya Sen, a leading economist and ethicist, views economics through a wide lens: "Economics is not solely concerned with income and wealth but also with using these resources as means to significant ends,

21

including the enjoyment of long and worthwhile lives."[1] However, the structures of economic globalization reduce that focus to income and wealth. As the critical component in the ecological crisis, these structures compromise the affluent in society and destroy the peoples and land they exploit.[2] North American theologian Larry Rasmussen talks about the totally different principles of the Big Economy (the globalized economy) and the Great Economy (the ecosphere or cosmic household of God).[3] Catharina Halkes, a noted European feminist theologian, dreams of an *oiko*-theology, attuned to the global reality, rather than an *ego*-theology, focused on the individual at the expense of the communal and cosmic context.[4]

This chapter begins with an overview of the global scenario, with specific reference to the impacts of globalization on our environment. This is followed by a critique that suggests that globalization is part of a mindset that has dominated the West in recent centuries. I suggest that we are at a new moment of decision, with an opportunity to shape a future with alternative values. Integral to this future is an environmental consciousness that appears to foster a hunger for new meaning. An engagement with our own spiritual tradition can equip Christians to work with others to encourage more holistic ways of relating to nature and one another.

The Global Scenario

The term "globalization" entered our vocabulary to describe the process whereby national economic structures are subsumed within a global system of finance and markets. Rasmussen names the nub of the problem when he says that this global economy "hops national frontiers with ease and generally ignores an international system that depended on territorial, sovereign and independent nation-states for its stability."[5]

Globalization produces a certain paralysis, partly because the global economic system is characterized by "placeless power and powerless places."[6] In democratic countries we are used to political leaders in places

1. Amartya Sen, "The Economics of Life and Death," *Scientific American* (May 1993) 18.

2. That crisis has well been called "ecocide" and is given a thumbnail sketch in Elizabeth A. Johnson, *Women, Earth and Creator Spirit* (Mahwah, N.Y.: Paulist Press, 1993) 5–9.

3. Larry L. Rasmussen, *Earth Community, Earth Ethics* (Maryknoll, N.Y.: Orbis, 1996) 111–126.

4. Catharina J. M. Halkes, *New Creation: Christian Feminism and the Renewal of the Earth,* 1st American ed. (Louisville: Westminister/John Knox Press, 1991) 156.

5. Rasmussen, *Earth Community, Earth Ethics,* 2.

6. Manuel Castells and Jeffrey Henderson, "Techno-Economic Restructuring, Socio-Political Processes and Spatial Transformation: A Global Perspective," *Global Restructuring and Territorial Development,* ed. Jeffrey Henderson and Manuel Castells (Beverly Hills, Calif.: Sage, 1987) 7.

we can name, elected by processes in which we have some say. Transnational companies are faceless, placeless, and accountable to no one other than shareholders. Their businesses straddle banks, media, agriculture, and the military. Such companies speculate in finance markets, moving billions of dollars per day, more than many national annual budgets. Futurist Hazel Henderson and others have named this untaxed trade a "global casino," where a few players hold whole nations to ransom.[7] Transnational companies push for new agreements so that they can claim compensation if national laws limit their profits.[8] They are beyond the control of any one country, and no international power has been given teeth to limit their activities.[9] No one, least of all local political leaders, can keep abreast of their plans, let alone restrain them.

Henderson identifies a number of interactive and accelerating globalizing trends: industrialism and technology; work and migration; finance; human effects on the biosphere; militarism and arms trafficking; and communications and planetary culture.[10] A gender perspective would indicate that in each of these, poor women and families are disempowered.

Economic globalization is the result of very deliberate policies imposed on countries as a prerequisite for participation in world trade. This globalization came from an assumption that what was good for the United States was good for the world. In every economic crisis for the last twenty years, leading academic pundits and international financial institutions insisted on the same strategies. A writer for *Le Monde Diplomatique* put the problem sharply: "Equipped with . . . a toolbox containing nothing more

7. In 1997 worldwide currency transactions averaged over US $1,000,000 million per day. See Anwar Ibrahim, "Globalisation and the Cultural Re-Empowerment of Asia," *Globalisation: The Perspectives and Experiences of Religious Traditions of Asia Pacific*, ed. Joseph Camilleri and Chandra Muzaffar (Malaysia: International Movement for a Just World, 1998) 6. The Tobin tax, originally suggested at .1% but now more realistically at .001 to .003%, on currency speculation (which is estimated at 93% of international currency transactions) has been resisted by banks, etc., and so is not pushed by national governments. See Mahbug ul Huq, Inge Kaul, and Isabelle Grunberg, eds., *The Tobin Tax: Coping with Financial Volatility* (New York: Oxford University Press, 1996) and Hazel Henderson, *Building a Win-Win World: Life Beyond Global Economic Warfare* (San Francisco: Berrett-Koehler Publishers, 1996) 318–319.

8. The Multilateral Agreement on Investments (MAI), which transnational companies tried to force on GATT and NAFTA, sought compensation for all national laws that reduced corporate profits. Action by concerned groups prevented this from being included, but the struggle continues.

9. The World Bank and International Monetary Fund are the two key financial institutions put in place after the Bretton Woods agreement with the aim of ensuring financial stability across the globe. They are regulatory bodies, responding to dominant economic and financial interests.

10. Henderson, *Building a Win-Win World*, 11–12.

than four large hammers (deregulation, privatization, tax reductions and free trade), the international economic organisations set about transforming the world according to the American model."[11]

Deregulation benefits the affluent, exploiting labor and resources. Privatization ranks profits for shareholders above investment in the future of the community. Tax reduction entails cuts to community services, including health and education. "Free" trade destroys local industries and skills bases. Cumulatively, these changes often result in poverty and powerlessness, leading to civil unrest and a consequent increase in militarization. Economic rationalism, spread on the understanding that it was value free, in fact assaults the culture, society, and livelihood of communities across the globe.

There are undisputed good effects of globalization, such as the spread of consciousness of human rights and access of poorer nations through the Internet to the top library resources of the world.[12] However, because international law has been developed most strongly in economic areas, economic priorities control other areas of global interaction.

The globalized economy affects the environment directly. Companies intent on profit permit practices that result in oil, chemical, and nuclear "accidents."[13] Arms manufacturers and dealers facilitate the destruction caused by landmines and mass bombing. Governments risk widespread contamination with nuclear and chemical weapons. Business interests manipulate the gene pool of seeds, seeking control of the food chain for profit, limiting plant diversity, and threatening havoc in communities that keep grain for future seeding. Poverty caused by debt repayment drives rural populations into hitherto untouched forests, scrounging for food, firewood, and building materials.

"Growth is good!" Globalization is shaped by economic policies based on the belief that growth, sustained by increased consumerism and measured by a higher Gross Domestic Product (GDP), is good. However, the GDP measures the growth of profits, without any counterbalance of

---

11. Serge Halimi, "From Market Madness to Recession: Liberal Dogma Shipwrecked," *Le Monde Diplomatique* (October 1998). In Australia this model has been called economic rationalism.

12. See Chandra Muzaffar "Globalisation and Religion: Some Reflections," in *Globalisation: The Perspectives and Experiences of Religious Traditions of Asia Pacific*, 179–190.

13. In Nigeria, Shell Oil has had disastrous effects on the land and community life of the Ogoni people. Nearer home in Ok-Tedi, Papua New Guinea, the Australian giant BHP's mining excavations turned the previously life-giving Fly River to sludge. Even in Australia the government has given permission for a uranium mine in Kakadu National Park, despite condemnation by World Heritage. Facts on these are readily available on the Internet or through social justice networks.

the common good. It does not measure environmental degradation or assign value to sustainable practices.[14] The adverse effects of mainstream economic policies have been felt in all nations.[15] Despite promises of trickle-down wealth, this fixation on economic growth, as Rasmussen points out, has led to "joint ecological and socio-economic impoverishment, at least at the periphery, where an increasing majority live."[16]

At its most basic, the logic of present trade and market patterns fails to ask about the requirements of environment, culture, and social relations for ongoing existence. Economic policies still work on the basis that the earth is flat and endless rather than closed and round.[17] David Hallman confirms that most measures ignore the fact that "merely tinkering with environmental regulations without a major restructuring of international economic relations is fundamentally flawed both practically and ethically."[18]

A sustainable lifestyle implies radical change in attitudes and practices. *Sustainablility* emphasizes the capacity of natural and social systems to survive and thrive together indefinitely.[19] Humanity is placed within the ecosphere as a key but interconnected participant rather than being understood as the sole referent in determining future policies. In a limited planet, unlimited growth is impossible. Ecological economists, developing alternative measures of societal health, insist that economic growth cannot be the only yardstick.[20] Wealth accumulation is only one indicator of social well-being. These economists take into account environmental debits associated with unsustainable practices. The affluent world, which consumes so large a share of the earth's resources and creates so much pollution, is slow

14. The Australia Institute, *Briefing Paper,* no. 12 (September 1997).

15. Michel Chussudovsky, *The Globalisation of Poverty: Impacts of IMF and World Bank Reforms* (London: Zed Books, 1997). This author presents numerous case studies to show the pattern and terrible social impact of globalization in its present form, with its present power structure.

16. Rasmussen, *Earth Community, Earth Ethics,* 125.

17. Ibid.

18. David Hallman, "Ethics and Sustainable Development," *Ecotheology: Voices From South and North* (Maryknoll, N.Y.: Orbis, 1994) 269–270.

19. Rasmussen, *Earth Community, Earth Ethics,* 127.

20. E.g., Manfred Max-Neef, *Human Scale Development: Conception, Application and Further Reflections* (New York and London: Apex, 1991); John Cobb and Herman Daly, *For the Common Good: Redirecting the Economy Towards Community, the Environment and a Sustainable Future* (Boston: Beacon Press, 1989); Clive Hamilton, *After Economic Rationalism: Rebuilding Communities* and other publications from The Australia Institute in Canberra. In *A Truly Civil Society* (Sydney: Australian Broadcasting Corporation, 1995), Eva Cox develops this fuller notion of societal well-being in an Australian context. Many of these authors suggest a new Genuine Progress Indicator or Index of Sustainable Welfare to measure a wider range of social "goods" than the market economy. See also Henderson, *Building a Win-Win World.*

to accept that the only way to achieve sustainable communities across the globe is to change deep-seated attitudes and practices. We are the ones with a debt: we owe it to the rest of the world to develop creative personal, national, and global policies to reduce consumption, to recycle, to conserve energy and prevent pollution.

Critiquing the Modern Mindset

The inequities of the present system impinge on all peoples, not only the two-thirds of the world who live below the poverty line. The environment and most of the world's inhabitants stand in a powerless place. Why have we allowed the earth and its people to be so harmed? Is it gross carelessness or willful choice? Perhaps it is neither. Despite inevitable human greed and desire for power, this situation is structural, not merely a result of personal unjust or inhumane choices. Economic rationalism in its globalized expression is a result of recent centuries where questions of value were separated from and overshadowed by choices that reflect technical expertise. Market possibilities were allowed to control society rather than society insisting that the economy serve the community.

This ideology is an end product of the Enlightenment, a movement of ideas that gradually reshaped European politics and society in the seventeenth and eighteenth centuries. In many seemingly distinct areas, ideas associated with the Enlightenment and the subsequent industrialization of society changed attitudes and structures in the Western world. The full impact came at the end of the twentieth century as inherited supports for community and spirituality eroded.

Basically, a guiding principle for the Enlightenment was the belief that the autonomous, individual, white male of financial means, independently of the community in which he lived, could apply critical reasoning to any and every problem without reference to traditional wisdom or authority. This scientific starting point affected religious, philosophical, moral, and political domains and, teamed with developments in the industrial revolution, it transformed horizons of thought, social organization, and self-understanding. Priority moved from the community to the individual; truth from dogmatic assertion to the empirically verifiable; the key referent from the divine to the human subject; obedience and tradition to autonomous choice. Fact and value were decoupled in philosophical thinking, and value was relegated to the personal and private domain. Social organization moved from oligarchic, aristocratic, religious control to new combinations of forces that based their "rights" in contract theories. Individual freedom grew, but communal responsibility was a casualty of technological expertise.

As a result of the Reformation a century earlier, the thin veneer of "Christian" Europe had blown apart. Families and kingdoms split, often using religious allegiance as a wedge. The resulting suspicion of religion as a divisive force, coupled with the increasingly obvious problem that what Christianity taught could not be proven as scientific fact, meant that religion was less acceptable in the public arena. New nations were formed in a consciously secular manner. Gradually the virtues of Christian teaching if not practice (compassion, care for the needy, honesty, fidelity) were relegated to the private arena or promoted as "charity." *Justice* became administration of law rather than expressing right relationships dependent on God. Market forces shaped the public arena, tempered in some nations by a welfare system that sought to provide a safety net for the less fortunate.

Awareness of the negative trends embedded in the modern paradigm should not obscure the contributions from this period. The person is recognized as critically important. Belief in freedom challenges structures of racial and gender privilege. Education is a goal for all. Even when breached, human rights are still recognized. The United Nations, though fragile, stands as a witness to the hope that cooperation between nations can triumph.[21] Democracy, while often exposed as a frail instrument, is nonetheless the best way that humanity has yet found to prevent tyranny. Science has rid our world of many diseases and offers hope even in cases where our present understanding is limited. Tolerance is accepted by all but the most fundamentalist religious groups.

Until recently, modernization and progress were considered synonymous. Western "advances" were proposed as the desired future for those they had colonized. However, many are noting that the cumulative effect of progress can be dysfunctional. Marc Luyckx suggests that people are discovering that "rational planning can take us efficiently to where we don't want to be when we get there."[22] In the West, communities are fragmenting. This fragmentation is embedded in our mobile lifestyles, our high walls, and houses sealed to the near neighbor while open to the world via the safety of the Internet. Western democracies struggle against apathy and passivity. Non-industrialized nations now question the vision they have been handed: while they may have become secular and materialist, they have not been enriched. Rediscovering their particular religious traditions, many adopt an intransigent, fundamentalist version to resist the encroachments of the West.[23]

21. Henderson, *Building a Win-Win World*, 293–303.

22. Marc Luyckx, "Religions and Governance: New Hypothesis," draft article for *Futures* (Brussels, 1999) 2.

23. Ibid., 4.

While the Churches raised questions about the secularization of society, this very trend ensured that their questions went unheeded. Church resistance to the modern was trapped in an agrarian worldview and language rather than challenging modernism on its own terms. Religion settled for the pews rather than the marketplace as its prime location and often attempted to prove its credibility by emphasizing its rational supports, isolating the experience of sacrament and mystery to the in-church world of incense and candles. By and large, the Churches unwittingly accepted the confines of the private domain, limiting ethical considerations to family matters and sexual mores, with only occasional, usually ignored, forays into the wider social arena.

The scientific approach removed limits to inquiry, exposing false religious constraints. However, the decline of religion as a moral force in the public arena happened at the same time as the rise of individualism. Given the competitive economic system, geared toward an ever-increasing market share, and the breakdown of village life, exploitation was rife in the growing urban sprawls. There was little discussion of the common good, little support for serious reflection on the directions being taken. *Realpolitik* unashamedly based itself on profit, competitiveness, and power. Successive crises showed that the poor, at home or in the colonies, were expendable.

Critical social theory, as developed by the Frankfurt school and later by Jürgen Habermas, raises serious questions about the modern era without wanting to return to the classical model.[24] Its analysis suggested that one of the driving forces behind modernity was the move toward a technical and instrumental mindset.[25] In the premodern era actions were evaluated by moral norms, by questions of purpose in relation to the larger picture. In the modern era the scientific revolution changed the main question from *why* to *how*. This approach found answers by researching ever smaller aspects of the whole. With scientific proof, agreement could be reached on how nature worked and how to change it. Without recourse to divine authority, it is hard to reach agreement on what is moral, so such questions were sidelined. A teleological evaluation gave way to an instrumental approach.

Habermas suggests that money and power, the values of the organized structures of commerce and politics, are so strong that they colonize the life-world—that world of family, relationships, attitudes.[26] The divide

24. See Paul Lakeland, *Theology and Critical Theory: The Discourse of the Church* (Nashville: Abingdon Press, 1990).

25. Steven Seidman, ed., "Introduction," in *Jürgen Habermas on Society and Politics: A Reader* (Boston: Beacon Press, 1989) 5. See James McEvoy's essay in Chapter 11 of this volume.

26. *Autonomy and Solidarity: Interviews with Jürgen Habermas,* ed. Peter Dews (London: Verso, 1992) 262.

between public and private is breached, not by compassion and love from the private sector, but by the "anti-values" of the public arena. Religion, nature, and women, confined to the private world, have been disregarded. However, campaigners, often inspired by their Christian faith, did force governments to put in place safeguards that ended slavery, required minimum standards in mines and factories, ensured basic education and health care, and eventually provided social welfare. Unfortunately the environment was not high on even the reformers' agenda until the latter half of the twentieth century.

Following is a simplified overview of these trends, which form the backdrop for global economic rationalism (see Table One).[27] The move from the agrarian to the modern mindset was so comprehensive that many call it a *paradigm* shift.[28] We are in the midst of another such paradigm shift. Much of the world is still agrarian, and the modern mindset still informs our structures and organizations, so the tension between these three paradigms is current. The third column indicates significant choices facing our time.

Seeds of Hope

How can we carry forward positive gains from the modern era while embedding them in a more interdependent value system? The third column suggests a period of immense change. A stark choice faces us: Will we support the forces working toward interdependence or stay trapped in former mindsets? Observers are calling this a "*kairos* time," characterized by danger and opportunity. If we fail to recognize the new moment, we are likely to head into devastation. But that is not the only possible direction.

Across the globe people are uneasy with values integral to the modern scenario. Some carry a nagging guilt that the global economy is unfairly skewed to benefit the affluent. Others are confused that our governments advocate international trade agreements like the General Agreement on Trade and Tariffs (GATT), which lead to unemployment of local workers. Others question the fairness of a system that allows big trade barons to avoid taxes while placing a regressive Goods and Services

27. This table is a variant of a table that summed up a series of insights from critical social theory and feminist and liberation theology in Christine E. Burke, *Credible Christian Participation in Public Discourse: Towards a More Just Society,* Ph.D. thesis (Melbourne: Monash University, 1995). It has been slightly modified after reference to the work of Marc Luyckx and the symposium he convened in Brussels.

28. A paradigm is like a pair of glasses through which we view reality. It constitutes an invisible frame of reference, of which the owner is not conscious. Paradigm shift is often used to describe a change that shakes the major pillars shaping our thinking and worldview. Such a shift envelops society only rarely.

TABLE ONE	Agrarian mindset	Modern, Enlightenment, industrial mindset (1750s onward)	Emerging (transmodern) mindset: 1990s into this millennium
A person's understanding of self in relation to others and society	• Determined by one's place in the religious, patriarchal, and hierarchical society. • Religion infused all of life. • The feudal lord was the reference point for belonging.	• Philosophers pursued a notion of self, based in rationality and individual freedom. This bypassed traditional wisdom or authority. • Gradual articulation of human rights accompanied increased access to education. Spiritual growth was less of a focus. • The nation became the significant reference point for belonging. Gradually, the limits to freedom based on class, race, and gender were exposed.	• Values personal freedom shaped by an interdependence that recognizes we are body-persons, needing a mutuality that respects rather than overrides difference. • Global and local networks are significant; spirituality is rediscovered as integral to humanity. OR • *Increasingly fragmented, alienated, and isolated; dominated by technology, degenerating into tribal conflict.*
The dominant understandings of Truth and Knowledge	• Revealed by God, giving meaning to whole of life. • Taught by religious authorities and not to be challenged.	• Rational, discovered and tested in scientific, objective, universal laws and explanations. • Non-rational ways of knowing were devalued because unscientific. Emotion and intuition were dismissed. • Moral values fragmented because the public arena lacked a common base. Increasing suspicion of dogmatic pronouncements. • Religion and women were relegated to the private sphere.	• Truth is a complex and multi-faceted goal to be sought, explored; experienced as partial, contextual, interpreted; revealed in story and symbol as well as scientific proof. • Overarching explanations and universal claims are distrusted. Knowledge recognised as a form of power, and critiqued as such. OR • *Deconstructed toward total fragmentation and relativity.* OR • *Invoked via a return to fundamentalist "certainties" that allow little or no dialogue with other perspectives.*

Worldview	• God-centred, ordered, coherent, static. • An "enchanted" vision, where events carried deeper meaning. • Shaped by Greek dualism, which devalued women and nature.	• "Disenchanted," *man*-centred, scientific; progress oriented; reflected the perspective of the economically powerful. • Dualism increases, rationality replacing spirituality as the highest good. Symbolically, humans and nature seen as machines with replaceable parts. • Competitive market believed to be "value free." • Nature there to serve human interests.	• "Re-enchanted," holistic, cosmic; aware of environment, wanting to preserve diversity. • Technology/communication temper growth to ecological needs, serving the just demands of economically disadvantaged two-thirds of world. • Aware of the organic interconnections of planet; listening to powerless. OR • *Continuing modern mindset, homogenized by global economy, media, and communications; self-destructing via pollution, consumerism, military might.*
Societal Organization	• Traditionally patriarchal; status based on birth, gender, marriage. • Shaped by village confines and past practices. • Religion had some power over politics.	• Increasingly urbanized national states develop with democratically elected governments; industrialized, geared to big business; shaped increasingly by media, technology. Growth of legal, educational, financial, business, and military institutions and government bureaucracies. • Life is divided into public/private spheres, with a breakdown of community cohesion. • In twentieth century human rights are clarified and claimed globally. UN struggles to contain national and economic divisions.	• Movements urge more *participative* democracy, restoring involvement and trust, seeking the common good; changes in availability and role of work. • Interdependence is developed via local or global networks; reconciliation becomes a critical issue. • Increasing recognition of the impact of colonial dispossession and conquest; NGOs grow in importance. • Women and minorities claim equal value and importance of difference. OR • *Societies characterized by increasing anonymity, economic and ethnic division. A permanent underclass emerging in even the most advanced industrial nations with accompanying civil repression and unrest.*

TABLE ONE continued	Agrarian mindset	Modern, Enlightenment, industrial mindset (1750s onward)	Emerging (transmodern) mindset: 1990s into this millennium
Nature	• God presented as bringing order out of chaos. • Humans charged with subduing the earth, which was awe-inspiring and beyond human control. • Nature a source of healing.	• Galileo and Newton discover laws "governing" nature. • To combat rationalism, in Western culture, nature was romanticized in art and music, yet ruthlessly exploited, "raped," as a resource for industry and agriculture. • Scientific research develops unimagined cures. • Indigenous people, with their deep collaboration with nature, are dismissed as primitive.	• Finite limits and the infinite value of nature are increasingly recognized, together with values implicit in indigenous peoples' relations with the land. • Nature respected as having an integrity that humans destroy at their own risk; biodiversity valued, solar power and alternative technologies are developed. • Movement toward sustainable lifestyles. OR • *Destruction via pollution, war, and exploitation. Science reshapes nature for profit via genetic engineering.*
Religion	• Central to social as well as personal life, giving meaning and shaping celebrations. • Priest seen as gateway to meaning and mystery, etc. • Church organization parallels feudalism and struggles with state power.	• Side-lined, part of private world, devalued. • Christianity divided, defensive; theology's rational emphasis increasingly remote from popular devotion. • Church community weakens as religion is perceived as irrelevant. • Increasingly, in twentieth century, Church exposed as patriarchal in a democratic world. • At end of era, Christianity is recognized as one among many world religions.	• Minority groups reexamine, reinterpret, and reclaim religious sources and rearticulate symbols. • People seek spiritual meaning. • Christian Churches develop communities respecting local diversity, providing a forum for social engagement and spiritual support. • Interreligious dialogue valued as providing a source of shared ethical wisdom. OR • *Continuing patriarchy, centralization, and irrelevance.* OR • *Return to fundamentalist understanding of key texts, which are used against others within or beyond the particular religion.*

Tax (GST) on ordinary folk. Many are apprehensive about our environment, the quality of our civil life, and the rights of workers. The gap between rich and poor widens inexorably.[29] Such feelings can foster a move from passivity toward action. Recent Non-Government Organization (NGO) demands for the cancellation of Third World debt gained support because people recognize the inequity of the system.[30] Naming the dislocation caused by this rationalist framework has encouraged new coalitions, insights, and values to emerge.

From this mixture of good and ill a new consensus is emerging. Henderson sees the growing dis-ease with modern priorities as a harbinger of a new mindset cohering around a few central values: "cultural diversity, social justice, the rights of the poor and indigenous peoples, women, and all those marginalized by economism."[31] This new approach that values "partnership, rather than the dominator paradigm of human organization, is being recognized by many as a safer evolutionary path."[32] Habermas also explores the shift to relationality raised elsewhere in this book.[33] He sees the challenge facing our age as a move from individualism to significant communication. Real knowledge comes, he suggests, not in passive acceptance of a *diktat* from on high nor in the solitary thinker surveying and manipulating the objective world, but in the process of communication between people who speak and act and seek to understand one another.[34] He suggests that communities of public discourse are needed if a consensus is to emerge to guide our future directions. This public discourse would encourage democracy to become more participatory and force us to forge some common moral bases for public action.[35]

Two recent seminars on "Governance and Civilizations" brought together participants from many nations to consider the hypothesis that the West could be in a process of transition from modernity to "transmodernity," a term coined to include keeping the best of modernity but going beyond it.[36] These seminars argued for an end to the separation between ethics and religion on the one hand, and politics and public affairs on the other, while insisting that the distinction between such areas be maintained.

---

29. In 1994 the top 10% of households held 50% of all household wealth, while the bottom 50% held only 3% of all household wealth. Figures from the Brotherhood of St. Laurence, quoted in "Poverty Watch," *Outlook* (April 1999).

30. Susan George, "Reverse Charges," *The Word* (1995) 3–5.

31. Henderson, *Building a Win-Win World*, 185.

32. Ibid., 193.

33. See Chapter 11 by James McEvoy in this volume.

34. Lakeland, *Theology and Critical Theory*, 39–69.

35. See *Autonomy and Solidarity: Interviews with Jurgen Habermas*, 248

36. Marc Luyckx, "Religions and Governance," 1.

The transmodern way of thinking, indicated in the first alternative in the third column of the table above, features a creative mix of rational and intuitive brainwork. Celebrating diversity, people who have moved to this approach believe that protection of the environment must be a central concern for every human being. They recognize ethics as a guide for private behavior and public policy, and move away from vertical "expert" authority systems toward flatter, more horizontal organizations and toward consensual decision making. A new interest in spirituality and a desire for a more sustainable relationship with the environment are integral to attitudes forming in this new moment.[37] A survey done in the United States indicated that two-thirds of the growing category of people who approach life in this new way are women. This ecofeminist link was manifest in the United Nations Women's Conference in Beijing, where women of all cultures committed themselves to environmental concern and justice for the most powerless.[38]

A variety of sources agree that opportunity beckons through this valuing of mutuality, participation, and interdependence, and openness to difference and minority voices. A thirst for spirituality is frequently linked to a readiness to see the wisdom in various traditions. A richer notion of truth, including a sympathy toward symbol and narrative, suggests a greater humility and a readiness to appreciate mystery. A holistic, organically connected, sustainable vision of the human in relation to the planet and the cosmos is replacing a careless superiority to nature. Many of these movements are kept alive by local action, taken in the context of global awareness.

A Christian Engagement in This Moment

Theologian Rebecca Chopp claims that Christianity is not merely a corrective but aims to bring the good news that anticipates and transforms the world.[39] If that is so, what are the implications of this moment for Christians concerned to live out their beliefs? Three connections spring to mind. This transmodern option provides a new context, a renewed call to conversion, and an opportunity for a specific contribution.

A new context: The transmodern option provides a new context that is surprisingly sympathetic to the heart of the Christian message and major world religious traditions. Christians do not have to instigate ecological concern; indeed, we are often latecomers. Globalization has brought the

37. Ibid.

38. Beijing Declaration.

39. Rebecca Chopp, *The Power to Speak: Feminism, Language and God* (New York: Crossroad, 1989) 18.

world closer together. Suffering beamed in from distant places can lead to a sense of our shared humanity, the breakdown of prejudice, and the valuing of other approaches. New technology can allow new solutions, such as solar power, recycling procedures, and pollution emission controls. Grassroots people in NGOs commit enormous effort toward a better world. *Religion, spirituality, reconciliation, business ethics, community, the common good*—all these concepts are integral to public discourse today. In all these, ecological concern is gaining ever more central ground.

However, turning our mindset toward the transmodern will be a long haul. Such a shift is not inevitable. The forces of individualism and consumerism are strongly entrenched and can deflect the yearning for spirituality toward a solipsistic "feel good" version, a McDonald's type of spirituality that Bonhoeffer might have called "cheap grace."[40] To keep going in the face of real adversity, to keep working in hope as yet another forest is cut down, will require a much deeper spirituality.

A renewed call to conversion: The transmodern mindset gives a new edge to the age-old call to conversion. Henderson says that global economics "has enshrined some of our most unattractive pre-dispositions: material acquisitiveness, competition, gluttony, pride, selfishness, short-sightedness and greed."[41] St. Paul could have written this in the first century rather than a world-renowned futurist at the end of the twentieth! The capitalist market economy has writ large what Pope John Paul II calls "the social structures of sin," the ways we allow systems and organizations to foster, encourage, and ultimately require these sinful tendencies of the heart.[42] This moment calls us from individualism to developing community, from power to participation and partnership, from domination to respectful interdependence with nature, and from words to deeds.

The modern economic system both indicates the need for redemption and opens us to recognize that need. Uncertainty has replaced an earlier arrogance. *Metanoia,* or "conversion," is being experienced as a need. The continuing outbreaks of ethnic warfare, with their impact on peoples and the earth, show that we cannot bring about change by political promises or military compulsion. Sustainable community "requires not just knowledge *(scientia)* but wisdom *(sapientia),* and the Psalmist's contrite heart and humble spirit."[43] Creation is indeed groaning for the redemption that is at

---

40. Mary C. Grey, *Beyond the Dark Night: A Way Forward for the Church?* (London: Cassell, 1997) 133.

41. Henderson, *Building a Win-Win World,* 88.

42. Pope John Paul II, *On Social Concerns (Sollicitudo rei socialis,* December 30, 1987) nos. 36–37.

43. Rasmussen, *Earth Community, Earth Ethics,* 167.

hand. Communities of hope are needed, offering a vision to sustain humanity as we seek to turn around our hearts and minds and expectations.

Environmentally respectful policies and attitudes require a civil society, in which the common good is not merely spoken of but put into practice.[44] Christian insights into social justice, along with those from other religious traditions, can bring a coherent understanding of human beings as persons in community, valuing the biosphere as an expression of God's creativity. If we could find a way of respectfully engaging in the civil discourse, the wisdom and humanity of many Christian social justice insights might deepen the civil discussion.

However, Christian Churches have been and are part of the problem. Such engagement in public discourse calls them to internal conversion in two key areas. First, feminist theologians have exposed a deep suspicion of matter, nature, women, and sexuality within the Christian tradition.[45] The resulting fear is far from Francis of Assisi's delight in nature. The exclusion of women from liturgical leadership and theological formation in many Churches has impoverished both spiritual theology and Church structures. Fear of what has been defined as "other" has driven a desire to control that has had a devastating impact on Christianity, a religion meant to value incarnation or "enfleshment."[46] This volume is part of the effort to reclaim a deeper tradition of respect and care.

The second challenge to us as Churches is to our own structures. Speaking with integrity in a world being torn apart by power and competition raises questions that need to be addressed within our own Churches. Leadership in movements is often diffuse, and people take initiative and action when and where they can. Authoritarian control is distrusted. Often Churches, within their own communities, embody domination. This power over others needs to be faced and repented of. Structures need to be changed. The present system limits credibility within the Churches and negates their ability to speak convincingly even to those in movements who share an ecological concern.[47]

An opportunity to contribute: What can Christianity contribute at this time of change? People are expressing a desire for spiritual sustenance. We bring our spiritual tradition. This will often be welcomed if shared

44. See Cox, *A Truly Civil Society.*

45. See Lucy Larkin's essay in Chapter 8 of this volume.

46. For example, Elizabeth A. Johnson, *Women, Earth and Creator Spirit,* and Anne Clifford, "When Being Human Becomes Truly Earthly," *In the Embrace of God: Feminist Approaches to Theological Anthropology* (Maryknoll, N.Y.: Orbis, 1995) 173–189. See also the following chapters in this volume: Chapter 4 by Duncan Reid, Chapter 5 by Patricia Fox, R.S.M., and Chapter 8 by Lucy Larkin.

47. Movements are rather unstructured gatherings of people who differ on many issues but are passionate for their central concern.

with a readiness to listen. The Churches can become more credible through participation in discourse and action that move society toward a new way of perceiving our place within the ecosphere. That requires familiarity with our own tradition, integrated in new responses to present questions. We also bring a history of active concern for those in distress. Movements are often characterized by the slogan "thinking globally, acting locally." This implies active commitment within a local community, while drawing on a larger vision of what is needed for humans to live in community with respect for creation. If we are to engage with the environmental movement, everyday Christians in our community must speak and act with integrity.

If we know our own tradition and draw richly from its wells, it is possible to bring hope into the dialogue. For this process we will need to listen to those scholars who reinterpret our tradition in Scripture and in systematic theology, as many essays in this volume show. Here I would like to tease out how this reappropriation could enrich our spirituality as we move toward ecological conversion.

Spirituality for the Long Haul

Over the centuries Christian faith has provided spiritual resources that have transformed communities. If we Christians would join with others in the movements for ecological healing, drawing on our spiritual resources, we might find a role supporting our brothers and sisters through the dark night of disillusion. The ecological reality challenges our spirituality to draw again from some of the central mysteries of our faith—Trinity, incarnation, and sacramentality—and also from the long tradition of discernment and reflectivity. Today women are integrating ecological concerns as they make important contributions to reclaiming and renewing theological spirituality. Their position on the periphery has enabled them to see new meanings in texts and practices previously interpreted from positions of power.

A spirituality that challenges the tenets of economic globalization would draw on the Trinitarian insights outlined in the chapters by Patricia Fox and Anthony Lowes in this volume. Mutuality, interdependence, and respect for difference find a symbolic expression in the core Christian belief in *God as Trinity*. An understanding of God as persons in dynamic communion places relationship at the heart of life, and this relating does not stop at the border of the human but stretches out to encompass all that expresses the life of God in creation.[48] Christians celebrate the value of

48. Denis Edwards, *Jesus the Wisdom of God: An Ecological Theology* (Homebush, N.S.W.: St. Pauls, 1995) 91–110.

loving interdependence, of mutual respect in diversity. This resonates strongly with the values of the transmodern era. This passion for relationship, valuing the person yet critiquing individualism, can be a gift to other groups that already see the threat to biodiversity and the pollution of air, land, and water, which sustain life. Valuing endangered species as expressions of God's magnificent diversity, as Denis Edwards's chapter suggests, can both strengthen commitment and be a source of hope. It can help us to see our concern as reflecting the concern of God and assure us that our efforts cohere with the saving action of One greater than us.

However, as noted above, for such a message to have any energy or to be experienced as carrying any "truth," it needs to come out of a lived base. Statements about a God of mutual love do not carry weight when coming from a patriarchal structure.[49] The very movement to this new appreciation of participation is closely allied with a distrust of unaccountable authority. A structure unable to appreciate difference within its own ranks, limiting all significant decisions and policy making to a peculiar subset and intentionally excluding a significant population because of "difference," cannot portray a God of inclusion and interdependence.

Incarnation is the mystery of God taking flesh, dwelling among us. The celebration of this mystery links the living, dying, and rising of Jesus Christ with the basic experiences of our lives. As pointed out in Duncan Reid's chapter, incarnation affirms the ultimate value of flesh, of this world, and thus counters the dualism that devalued the enfleshed, the here and now. Jesus' life bespeaks "acting locally, thinking globally." His life was so circumscribed by the particularity of race, place, and time, yet his impact has been global. This affirms the importance of seeing meaning in small daily efforts to think and act in ways that respect nature. Surprisingly, the recent work of scientists supports this sense that change can build up, moving from attitudinal shift to a new possibility. Margaret Wheatley, in her exciting *Leadership and the New Science,* writes: "Acting locally allows us to work with the movement and flow of simultaneous events within that small system. . . . These changes in small places, however, create large-systems change, not because they build one upon the other, but because they share in the unbroken wholeness that has united them all along."[50]

---

49. It is significant that women theologians are among the vanguard working to connect theology with the issues of ecology, echoing the predominant presence of women in the movements for ecology, peace, and reconciliation. See Robert J. Schreiter, *The Ministry of Reconciliation: Spirituality and Strategies* (Maryknoll, N.Y.: Orbis, 1998) 25–31.

50. Margaret J. Wheatley, *Leadership and the New Science: Learning About Organization from an Orderly Universe* (San Francisco: Berret-Koehler Pub., 1992) 42.

This mystery of Jesus' life, death, and resurrection can also give access to meaning for those who suffer. Many theologians are revisiting previous understandings of suffering and finding new insights in the Christian message.[51] Roberto S. Goizueta writes that the suffering poor, with their deep devotion to the suffering of Christ, express in traditional devotions their solidarity with him, who suffers for them:

> In the *Via Crucis*, Latinos and Latinas affirm the truth of the Resurrection not as an event that, subsequent to the Crucifixion, "overcomes" or "cancels out" the death of Jesus, but as the inextinguishable love and solidarity that defines the *Via Crucis* itself, as the act of accompaniment that constitutes and empowers us as persons and as a community of faith. It is precisely the intrinsically communal character of our confrontation with and struggle against death that already embodies the victory of life—of love—over death.[52]

Their reenactment of the death-resurrection shows that God is with them, and their celebration shows that they experience this as a community. The crucifixion does not explain, far less justify, the suffering, but it does assure com-passion. The risen Jesus is experienced in community. Their processions portray the importance of the person in community rather than as a lone individual. It is only in Christ-as-community, a community formed by the Holy Spirit, that this deeper understanding of the solidarity at the core of the death and resurrection of Jesus can be grasped. Goizueta writes:

> The very notion of communion can itself become abstract unless mediated by a praxis of accompaniment, a praxis of compassion, a praxis of justice wherein the starting point and ground of all communion is the victim's invitation to "touch my wounds," wounds that resist all attempts to define the victim in isolation from his or her relationships to others. Only by responding to that invitation, acknowledging one's complicity, and accompanying the victim in his or her struggle, can an authentic reconciliation take place. This, indeed, is the "preferential option for the poor" that forms the basis of all true community.[53]

The cross points to a different understanding of power, one that operates from within, not over, others. It does not value being a victim but stands as a sign of struggle against oppression. Such a message of salvation

---

51. See Elizabeth A. Johnson, *She Who Is: The Mystery of God in Feminist Theological Discourse* (New York: Crossroad, 1992) 246–272, and Lorna Hallahan's essay in Chapter 6 of this volume.

52. Roberto S. Goizueta, "A Matter of Life and Death: Theological Anthropology Between Calvary and Galilee," *Catholic Theological Society of America: Proceedings* 53 (1998) 3.

53. Goizueta, "A Matter of Life and Death," 14.

through suffering can come only from those who have entered into or stood alongside suffering. If we are part of the earth, then we suffer with it. Again, Rasmussen says it well: "The only power that can truly heal and keep the creation is power that intimately knows its degradation . . . . This kind of power, learned in suffering, expressed as empathy and compassion, and deeply sensed and felt, is what wins space for joy and abundant life."[54]

Christians who live in the affluent sector of the world can learn this in silence only as they experience a little of what others face daily. When it is proclaimed and lived by those who do suffer, then it carries weight for the wider movement. As we come to see not only other humans but nature as an unrepeatable expression of God and recognize our intrinsic communion with this web of life, our solidarity demands this praxis response. This might be in protest, in action to reclaim abused land, in tree planting, or in a myriad of communal efforts to raise awareness and change our attitudes.

A third element in this spirituality is nature as *sacrament*. A great tradition within the Christian heritage holds that ordinary, everyday things alert us to the deeper reality of God's presence among us. Bread, wine, oil, water, human bodies—all speak of a greater mystery that shines through, and is encountered in, life. The word "sacrament" indicates this ability to point to a deeper meaning and presence of God.[55] St. Paul said, "Your body is a temple of the Holy Spirit, who is within" (1 Cor 6:19). Our bodies are where we connect with nature. We, like the whole created universe, are "made from stardust."[56] By seeing our intrinsic link to all beyond us, we can widen our human boundary to include the environment on which we depend.

Nature has always had this potential of pointing to the divine. Many are easily, if inchoately, aware of this mediating role of nature. Surfboard riders, bush walkers, artists, poets, and gardeners all recognize the sacredness of nature, though many would not connect this with their mainstream faith. In part this could be because the last few centuries have successfully separated liturgy from life and made what happens in the church the prime concern of many who minister full time in the church. The connection to the everyday is lost, and the spiritual moments that people experience in nature are more often named by New Age religions than seen as integral to Christian spirituality. Recognizing and exploring symbols that connect us with this fragile planet as our home can heighten our awareness of our interconnectedness with nature and ground our commitment to live sustainably in a sense of reverence.

54. Rasmussen, *Earth Community, Earth Ethics,* 287.

55. See Susan A. Ross, *Extravagant Affections: A Feminist Sacramental Theology* (New York: Continuum Publishing Company, 1998).

56. Denis Edwards, *Made from Stardust: Exploring the Place of Human Beings Within Creation* (Blackburn, Vic.: Collins Dove, 1992).

If the Eucharist and other liturgical rites foster this sacramental consciousness, then minds and hearts might more easily connect with the sacramental moments in everyday life.[57] In a recent article William T. Cavanaugh links globalization and the Eucharist.[58] He argues that globalization both detaches people from their local reality, as fast-food chains replicate dreary versions of local foods for global consumption, and "produces fragmented subjects unable to engage a catholic imagination of space and time."[59] Cavanaugh affirms the countercultural power of the Eucharist grounded in local community, yet conscious of global connectedness as the whole body of Christ.[60] Gathering all the faithful, both rich and poor, around the altar exposes the barriers that we put up, as St. Paul told the Corinthians (1 Cor 11:17-34). An awareness of our intrinsic connection to the fragile earth, which provides our food and drink, can heighten our appreciation of that for which we give thanks.

Around all these elements of spirituality is another tradition, a "how to" tradition. Reflective prayer and *discernment,* so that one knows when to make an issue the sticking point and when to let go and entrust it to Providence, are underutilized resources. A strong element in a number of Christian spiritual traditions is a daily process of reflection, discerning the action of the Spirit in the ordinary situations of life and recognizing the responsibility of significant choices. Recent developments in theological reflection help people to connect their faith and their daily lives in ways appropriate to our times.[61] Church communities can use these approaches to help members deepen their spiritual life.

This discipline of reflection on personal and communal choices could shape community and become part of the rhythm of daily life. If our connection with the environment were integral to such reflection, change could be grounded in a respect for creation. This is especially true for affluent sectors, where we consume and pollute far more than our numbers warrant. Being the carriers of such a spirituality suggests a conversion, not just to see the environment as a key issue, but to see our own tradition and organization with new eyes and to reshape our selves in line with the gospel.

57. See Chapter 1 by Stephen Downs in this volume.

58. William T. Cavanaugh, "The World in a Wafer: A Geography of the Eucharist as Resistance to Globalization," *Modern Theology* (April 1999) 181–194.

59. Ibid., 182.

60. Ibid., 193.

61. Patricia O'Connell Killen and John de Beer, *The Art of Theological Reflection* (New York: Crossroad, 1994).

Conclusion

Ecological concern requires present action with a future orientation. Jesus' life calls his followers to be countercultural when the values shown in his life require it. Those who care for the earth seek to hand on to future generations a place capable of sustaining and delighting them. If this passion can be fostered through the networks that the Churches access through prayer, reflection, and spirituality, then there is a real possibility of widening and deepening the groundswell toward active participation to prevent the destruction of our world.

For many communities this will look like stargazing. However, an increasing number of people questioning the focus on the economy seek new meaning, and they carry new energy.[62] As theological educators and students allow ecological awareness to shape our work, a more integrated awareness will be communicated in our ministry.

Any suggestions that the Christian tradition can play a supportive role in the move toward a transmodern society require that the Church itself change. Some new combination of participation with creative leadership is needed. A spiritual conversion on power issues is fundamental. A firm commitment by the Churches to environmental concern could nourish renewal at the grass roots and provide energy and hope for the world beyond the Churches. It could also strengthen the global networks of Churches in ways that unite participants, not just hierarchies.

I have argued that the globalization of the economy, which is a given in our world, has had serious impacts on the social and ecological environment. I have indicated that the roots of this malaise are deep-seated and that any response must be aware of these roots, which have dominated societal attitudes and expectations over recent centuries. In the final section I have suggested that the Christian Church is in a position to participate in turning around our individualistic mindset toward one that is holistic and inclusive. The whole world situation calls for humility, for openness to conversion, and for a way of acting that respects the gifts and insights of others. That sounds like a gospel-based agenda. Can we risk taking it?

---

62. Paul Ray, *Integral Culture: A Study of the Emergence of Transformational Values in America* (1997), quoted in Luyckx, "Religions and Governance," 3.

Goannas

If you keep very still,
They always wander by.
They're eye-catching.
To me, they're beautiful.
They have Personality.
They're special.
I feel a friendship.
They're part of our country.
Part of our dreaming.

Donna Bruce

Chapter 3

"For Your Immortal Spirit Is in All Things"

The Role of the Spirit in Creation

Denis Edwards

North of Adelaide there is an ancient and beautiful place called the Flinders Ranges. These ranges were formed from the eroded remnants of once massive mountains that were thrown up by movements in the crust of Earth about 500 million years ago. They are home to wallabies, kangaroos, emus, cockatoos, and eagles. Scattered through the ranges are engraving and painting sites that are sacred to the humans who inhabited the area, the Adnyamathanha people.

Recently I had a chance to return to this area for a few days. Like many other people, I find it a place in which it is easy to unwind, to let go of timetables and the demands of life, and to simply be present to the hills, trees, birds, and animals. The Flinders Ranges are restorative and healing and bring me to a sense of peace. My experience has always been that this is a place of grace, a place where the Spirit of God dwells, a place where the Spirit can be met if one takes the time to be there quietly.

This kind of experience, which seems to occur in the lives of many people in various places, raises fundamental theological questions: What is the connection between the Holy Spirit and such a place? Is the Holy Spirit really present and mediated through place and through the complex of mountains and gorges, trees and animals? The essays by Phillip Tolliday and Stephen Downs in this volume take up the issue of place in ecological theology. My focus here will be specifically on the presence of the Spirit in and to creatures. Is the Holy Spirit really present in all creatures? Is it in any way distinctively the work of the Spirit to be at the heart of creation?

Traveling through Brachina Gorge, one of the passes that run through the ranges, adds a further depth to these questions. It invites reflection not simply about the presence of the Spirit now but also about the Spirit's role in the evolution of life. It leads to questions about the Spirit's role not just with regard to *place* but also with regard to *time*. At one end of Brachina Gorge, there are stromatolites that are about 630 million years old. These are fossilized mats formed from the interaction of simple bacterial forms of life with sedimentary particles. They point back to the time when life took the form of single-celled, microscopic organisms living in the sea. These micro-organisms are related to the bacteria that emerged as the first forms of life about 3.9 billion years ago and to the bacteria that still flourish today.[1]

At the other end of Brachina Gorge, there are fossils from the early Cambrian period. In this period (from about 540 million years ago) multi-celled, complex organisms evolved and flourished in the seas. The abundance of life in the early Cambrian period has been described as the most dramatic explosion of diversity in evolutionary history. The Cambrian fauna of Brachina Gorge include trilobites, mollusks, and many examples of cup-shaped creatures called archaeocyaths.

Between these two sites, in the center of Brachina Gorge, there are rare fossils of some of the oldest known forms of animal life. These creatures, called the Ediacara Fauna, lived about 600 million years ago at the bottom of what was probably then a shallow tidal lagoon. They were soft-bodied, but they are preserved as impressions in quartzite. Among these impressions, there are some that appear to have been anchoring bulbs of frond-like creatures, others that seem to have been polyps attached to the sea floor, and others that were ancient worms as well as a variety of unknown creatures.[2] A journey through Brachina Gorge raises the question: What is the relation of the Holy Spirit to the evolution of life in all its manifestations, from the original bacteria to the developed creatures of the Ediacara, and the wonderful diversity of life in the Cambrian period?

I will begin by addressing the question: What, if anything, can be said theologically about the *distinctive* role of the Holy Spirit in ongoing creation? I will attempt to suggest an understanding of the Spirit's role that is consistent with the best insights of the Christian tradition, that is plausible in the light of contemporary evolutionary views, and that can serve as a

1. Lynn Margulis has argued that ancient bacteria have been incorporated symbiotically inside plant and animal cells as plastids and mitochondria and also that the first nucleated cells (but not their nuclei) have a symbiotic origin. She points out that while the origin of the nucleated cell remains controversial, the symbiotic origin of plastids and mitochondria is now widely accepted. See her *Symbiotic Planet: A New View of Evolution* (New York: Basic Books, 1998) 33–49.

2. Margulis tells us that these ancient fossils are now recorded at about twenty sites worldwide. She also comments that it is hard to know whether the Ediacara are properly animals or protoctists or a mixture of both. See *Symbiotic Planet,* 93, 95.

foundation for an ecological theology of the Spirit. However, there is a consistent theological tradition that claims that the actions of the Trinity toward creation are undivided and one, and this has often been taken to mean that there is no *proper* role for any one of the Trinitarian Persons in creation. In the light of this, I will ask in a second section: Can any such distinctive role be described as *proper* to the Holy Spirit? Finally, I will conclude with some reflections on the ecological consequences of this kind of theology of the Spirit.

The Distinctive Role of the Spirit in Creation

John Zizioulas distinguishes between two types of Spirit theology in the New Testament. In the first the Spirit is seen as the power *(dunamis)* for mission, while in the second the Spirit is seen as the eschatological gift of communion *(koinonia)*.[3] In what follows, I will take up this idea that the Spirit's role can be understood in terms of both empowering and communion-bringing. Where Zizioulas's focus is on the Spirit as the empowerer for mission and the bringer of communion in the creation of Church, I will suggest that the distinctive work of the Spirit in the ongoing creation of all things can be understood in terms of the power of becoming and the gift of divine communion with each creature.

The Life-Giving Spirit as the Power of Becoming

According to the common scientific story, our universe had its beginning twelve to fifteen billion years ago in an expansion from a state that was unthinkably small, dense, and hot. This process, called the Big Bang, is not to be understood as an explosion of matter in an already existing space and time but as an expansion and stretching of space-time itself. The high temperatures of the early universe produced a world that was made up of about 75 percent hydrogen, 25 percent helium, and tiny amounts of other elements. This matter was "clumpy" enough to allow galaxies to form as the universe expanded and cooled, and this meant that the universe could light up as stars ignited from compressed hydrogen in a process that is called nucleosynthesis. In this process other elements are "cooked"

---

3. Jean Zizioulas, "Implications ecclésiologiques de deux types de pneumatologie," *Communio Sanctorum: Mélanges offerts à Jean-Jaques von Allmen* (Paris: Labor et Fides, 1981) 141–154. Zizioulas shows that both approaches are dimensions of the New Testament understanding and that while the West has favored the missionary emphasis and the East that of communion, both dimensions are needed today for a genuine ecumenical theology of the Spirit and of the Church. The essays by Patricia Fox (Chapter 5) and Anthony Lowes (Chapter 7) in this volume explore important aspects of Zizioulas's theology.

from the original hydrogen, and in the more intense heat of supernovas, still heavier elements are forged. From these processes come all the elements that make up our Earth, with its kangaroos, kookaburras, and human beings.

About 4.6 billion years ago, the Sun and what would become our solar system began to form near the inner edge of one of the spiral arms of the Milky Way Galaxy. Earth took shape from accretions and collisions of nebula material. Over the next several hundred million years it developed a core, a crust, a primitive atmosphere, and an ocean. Life seems to have appeared on Earth about 3.9 billion years ago. Communities of bacteria formed the planet's first ecosystems, and the evidence from a number of disciplines leads to the conclusion that the whole complex pattern of life on Earth has evolved from these communities of simple cells. About 7 million years ago in Africa, an apelike species evolved that had an upright *(bipedal)* style of walking. *Homo erectus* seems to have emerged about 2 million years ago, with various archaic forms of *Homo* appearing between 500,000 and 34,000 years ago, and modern humans *(Homo sapiens)* about 120,000 years ago.[4]

It is the task of science to describe the story of the unfolding of the universe and the emergence of life, but it is the task of theology to ask about the relation of the Spirit of God to what is proposed in the scientific account. I want to suggest that it is congruent with the Christian tradition to see the Spirit as the power of becoming who enables the emergence of a life-bearing universe. The Spirit can be understood as the presence of God to things that enables them to be and to become.

A rich resource that can be invoked in support of this view is the Christian confession of the Holy Spirit as the "Giver of life."[5] In the Creed associated with the First Council of Constantinople (381), the words used to express the divinity of the Holy Spirit stay close to the language of the Scriptures. Following Paul (2 Cor 3:17), the Creed describes the Holy Spirit as "Lord." Following John's Gospel (John 6:63), it describes the Spirit as the "Life-giver."[6] The theological discussion of the Spirit that preceded the Council, particularly the work of Basil of Caesarea, made it clear

4. There is a debate about the emergence of modern humans, with some favoring the "Multiregional Hypothesis" and others supporting the more recent "Out of Africa" hypothesis. On this see Richard Leakey, *The Origin of Humankind* (London: Phoenix, 1994) and Chris Stringer and Robin McKie, *African Exodus: The Origins of Modern Humanity* (London: Pimlico, 1996).

5. For this reason, of course, the theme chosen for the Assembly of the World Council of Churches in Canberra, Australia, was "Come, Holy Spirit, Renew the Face of the Earth."

6. The Creed also describes the Holy Spirit as "proceeding" from the Father (John 15:26) and as "having spoken through the prophets" (2 Peter 1:21).

that the Spirit is to be understood as Life-giver not only in the work of sanctification of human beings but also in the work of creation of the world.[7]

In the New Testament, the emphasis in texts that use "life-giving" language of the Spirit is on the Spirit's role in our sanctification and participation in risen life (1 Cor 15:44-45; 2 Cor 3:6; Rom 5:5; 8:2; 8:9-11; John 3:3-8; 6:63). This is not surprising in communities that are close to the overwhelming events of Easter and Pentecost. But this New Testament usage depends upon the older biblical notion that *biological* life is a gift of the life-giving Spirit. At the beginning of the first creation account in the Book of Genesis, we are told that "a wind *(ruach)* from God swept over the face of the waters" (Gen 1:2). The same Hebrew word *ruach* can mean "wind," "breath," or "spirit." This wind that "sweeps" or "broods" over the waters of chaos is the sign and the forerunner of God's creative activity, which is initiated by God's word in the next verse: "Then God said, 'Let there be light'; and there was light" (Gen 1:3). Throughout the Book of Genesis, the breath of God is the principle of life for all living things. Creatures survive only because God gives them the *ruach*. God forms the first human from the dust of the ground and breathes into this earthling's nostrils "the breath of life" (Gen 2:7). It is this breath of God that keeps mortals alive for their allotted span (Gen 6:3). God threatens that the flood will "destroy from under heaven all flesh in which is the breath of life" (Gen 6:17). Those that are to be saved go into the ark with Noah, "two of all flesh in which there is the breath of life" (Gen 7:15).

In the Book of Job we find Elihu saying, "The Spirit of God has made me, and the breath of the Almighty gives me life" (Job 33:4). And a little later he declares of all living things, "If he should take back his spirit to himself, and gather to himself his breath, all flesh would perish together, and all mortals return to dust" (Job 34:14-15; cf. Eccl 12:7). In the great scene in Ezekiel, the valley of dry bones can only be brought to life by the breath from God. The prophet is instructed: "Prophesy to the breath, prophesy, mortal, and say to the breath: Thus says the Lord GOD: Come from the four winds, O breath, and breathe upon these slain, that they may live" (Ezek 37:9). In the Book of Judith there is the ancient prayer: "Let all creatures serve you, because you spoke, and they were made. You sent forth your spirit and it formed them" (Jdt 16:14). In the Book of Psalms too, the word of God and the breath of God are linked together as creative forces:

---

7. In an early work, *Against Eunomius,* which appeared about 364, Basil associated the Holy Spirit with the work of *life-giving* and with the work of *perfecting.* The breath of God brings things to life and brings them to completion. Basil appealed to Psalm 33:6 to show how the Spirit is at work in creation: "By the word of the Lord the heavens were made, and all their host by the *breath* of his mouth." In both arenas, creation and sanctification, the Holy Spirit is the Life-giver and the one who brings to wholeness.

> By the word of the Lord the heavens were made,
>   and all their host by the breath of his mouth" (Ps 33:6).

In Psalm 104, the great song of creation, we find the psalmist singing of God's creatures:

> . . . when you take away their breath,
>   they die and return to their dust.
> When you send forth your spirit, they are created;
>   and you renew the face of the ground (Ps 104:29-30).

The creedal and biblical affirmation of the Spirit as the Giver of life provides a solid foundation in the tradition for a Christian theology of the Spirit's work in ongoing creation. This is not yet, of course, an evolutionary theology. But I would suggest that the traditional notion of the Spirit as Life-giver is open to development in the direction of evolutionary theology. One way of moving in this direction is by building on an idea that appears in Karl Rahner's evolutionary Christology. Rahner's theology begins from the standard theological conviction that all creaturely existence is contingent upon God's continuing creation. Things exist only because God conserves them in their being and in their action. All creatures owe their existence at every moment to the ongoing creative activity of God.

Rahner points out that when we move to an evolutionary way of thinking, we have to account for the appearance of what is substantially new, in events such as the emergence of life and of self-conscious human creatures. He maintains that such evolutionary events have inner-worldly causes and that it is the work of science to explain those causes. But he insists that the advent of what is radically new in evolution needs accounting not only at the level of science but also at the deeper level of theological reflection on God's ongoing creative action.

How are we to understand God's ongoing creative role in the evolution of what is profoundly new? Rahner suggests the notion of "active self-transcendence."[8] By this he means that God can be understood as creatively present to evolving creatures, not simply enabling them to *exist* in a static way but enabling them, in certain circumstances, to go beyond what they already are. God enables them not only to be but also to become something that they were not. The power of self-transcendence comes from *within* creation itself, but it is a power that finally comes not from nature but from the ongoing creative activity of God. God upholds and empowers the process of evolution from within, as the power enabling

---

8. See, for example, his article, "Evolution: II. Theological," *Encyclopedia of Theology: A Concise Sacramentum Mundi*, ed. Karl Rahner (London: Burns and Oates, 1975) 478–488.

creation itself to bring about something new. God, then, is not understood as intervening as one cause among others but as the always-present, dynamic Creator enabling creatures not only to exist but also to transcend themselves and to become what is new.

Rahner has discussed this idea particularly in his evolutionary Christology.[9] He sees Jesus Christ as both the self-transcendence of the universe to God and God's absolute self-communication to creation. I find Rahner's thought rich and stimulating here and believe it can be developed, with the idea that *the process of self-transcendence can be seen as the distinctive work of the life-giving Spirit.* In the long tradition of Trinitarian theology, we find that it is precisely the role of the Spirit to be the one who is the principle of the indwelling of God in creatures. Athanasius encapsulated the Eastern patristic tradition when he said that "the Father creates and renews everything through the Word in the Spirit."[10] All divine action is "in" the Spirit. It is the Spirit who dwells in things. The triune God is immanent in all things in the Spirit. This immanence of the Spirit will be the theme that is taken up and developed in the next section.

I have been suggesting that it is entirely consistent with the biblical and theological tradition to understand the Spirit, in our day, as the Life-giver, who is the power of becoming at the heart of creation. As Walter Kasper puts it, the Holy Spirit is to be seen as the source of novelty in creation. He writes:

> Since the Spirit is divine love in Person, he is, first of all, the source of creation, for creation is the overflow of God's love and a participation in God's being. The Holy Spirit is the internal (in God) presupposition of communicability of God outside of himself. But the Spirit is also the source of movement and life in the created world. Whenever something new arises, whenever life is awakened and reality reaches ecstatically beyond itself, in all seeking and striving, in every ferment and birth, and even more in the beauty of creation, something of the being and activity of God's Spirit is manifested.[11]

With Kasper, I want to suggest that it is the life-giving Spirit who is the source of the new, the power that enables creatures to transcend themselves. It is the Life-giver who enables the movement of the unfolding of

9. See Karl Rahner, "Christology Within an Evolutionary View of the World," *Theological Investigations* 5 (London: Darton, Longman and Todd, 1966) 157–192, and his *Foundations of Christian Faith* (New York: Seabury Press, 1978) 178–203. See also his *Hominisation: The Evolutionary Origin of Man as a Theological Problem,* trans. W. T. O'Hara (New York: Herder and Herder, 1965).

10. *Letters to Serapion,* 1, 24, in *The Letters of Saint Athanasius Concerning the Holy Spirit,* trans. C.R.B. Shapland (London: Epworth Press, 1951).

11. Walter Kasper, *The God of Jesus Christ* (London: SCM, 1983) 227.

the early universe from the Big Bang, the beginning of nuclear processes in stars, the formation of our planetary system, the emergence of life on Earth, and the evolution of self-conscious human beings. Long ago, in her hymn on the Holy Spirit, Hildegard of Bingen sang:

> Holy Spirit, making life alive,
> Moving in all things, root of all creative being,
> Cleansing the cosmos of every impurity,
> Effacing guilt, anointing wounds,
> You are lustrous and praiseworthy life,
> You waken and re-awaken everything that is.[12]

Hildegard spoke of the vivifying effect of the Spirit with the word *viriditas,* a Latin noun that means "green" or "greenness." In her thinking, *viriditas* is associated with the freshness, fecundity, and vigor of life. It means "not only verdure or foliage, but all natural or spiritual life as quickened by the Holy Spirit."[13] Elizabeth Johnson builds on the tradition of Hildegard with her feminist theology of the Spirit at work in creation:

> The Spirit is the great, creative matrix who grounds and sustains the cosmos and attracts it toward the future. Throughout the vast sweep of cosmic and biological evolution she embraces the material root of existence and its endless new potential, empowering the cosmic process from within. The universe in turn is self-organizing and self-transcending, corresponding from the spiraling galaxies to the double helix of the DNA molecule to the dance of quickening power. The Spirit's action does not supplant that of creatures but works cooperatively in and through created action, random, ordered or free.[14]

It is this same Spirit who empowers the life and ministry of Jesus of Nazareth within our evolutionary and cultural history as the radically Spirit-filled human being. It is this same Life-giver who is the power of the new creation, the power of sanctification and transformation of all things in Christ. It is this Spirit of God who, always in profound communion with the other divine Persons, is the divine creative power of the being and becoming of the universe.

12. Matthew Fox, ed., *Hildegard of Bingen's Book of Divine Works with Letters and Songs* (Santa Fe: Bear, 1985) 373.

13. Barbara J. Newman, "Introduction," *Hildegard of Bingen: Scivias,* trans. Mother Columba Hart and Jane Bishop (New York: Paulist Press, 1990) 25.

14. Elizabeth A. Johnson, *Women, Earth, and Creator Spirit* (New York: Paulist Press, 1993) 57–58. In other essays in this volume, Patricia Fox discusses Elizabeth Johnson's work in detail (Chapter 5), and Lucy Larkin offers a quilt that brings together a number of other ecofeminist theologies (Chapter 8).

The Life-giving Spirit as Ecstatic Gift of Communion

In an attempt to articulate the role of the Holy Spirit in creation, I have been suggesting that the Spirit can be understood as the Life-Giver who empowers the evolutionary becoming of creation. But this is only an initial insight into the Spirit's role with creatures. The Spirit's role with each creature needs to be understood as a relational and Personal one. The Holy Spirit creates a relation between each creature and the divine perichoretic communion. It is this relation that enables a creature to be and to become. The Spirit is not to be understood simply as an impersonal *power* but as a *personal presence* interior to each creature, creating communion with all creatures in ways that are appropriate for each of them. It is the presence of the Spirit that enables creatures to interact in their own creaturely patterns of relationship, at the level of protons, cells, evolutionary symbiosis, ecological niches, the planetary community, the solar system, the Milky Way Galaxy, and the universe.

In arguing for a personal divine presence to all creatures, I do not mean to suggest that the divine presence to a granite boulder is the same as the divine presence to a human being by grace. In both cases the presence is personal on the side of God. But it is a personal presence that relates to granite precisely as granite and that relates to a human being as a grace-filled, self-conscious, dialogical other. We humans are capable of a personal engagement with other creatures that takes delight in them, feeling with them in their struggle and pain, valuing them in their own right for what they are, and respecting them in their specificity and in their interrelationships in an ecological whole. This experience, by way of analogy, enables us to glimpse the differentiated personal presence of the Spirit to each creature, a presence that creates a bond of communion with each, a communion that empowers its being and becoming.

I want to suggest that these two dimensions, *dunamis* and *koinonia,* are interrelated with each other in the *one* presence of the Spirit to each creature. They are distinguished only in order to bring out the fullness of the Spirit's role. The Spirit is the one who *empowers and creates* precisely as the one who *relates* to each creature, bringing each into communion with the Trinity. This empowering personal presence unites each creature in a world of relations with all other creatures in a profoundly relational universe. This communion between each creature and the divine Persons-in-communion *is* the relation of ongoing creation.

At the heart of Christian faith is the conviction that the God who radically transcends creation is also intimately present to every creature. Augustine speaks of the divine presence as *interior intimo meo,* that is, as closer to me than I am to myself.[15] While Christian theology understands

---

15. See *Confessions,* III, 6, 11.

God as present to human creatures in a distinct, gratuitous, and interpersonal way through the gift of grace, it also points to this same God as
already present interiorly to every creature by the relationship of ongoing
creation. Traditionally, theology has used the language of divine omnipresence and of divine immensity to describe God's presence to all of
creation. God is interiorly and intimately present in all that God creates.
This tradition of the divine interior presence can be enriched when it is
connected to the idea that it is precisely the role of the ecstatic Spirit, in
koinonia with the other Trinitarian Persons, to indwell in all creatures. In
the Bible the divine interior presence is related to the life-giving Spirit in
Psalm 139:7-10:

> Where can I go from your spirit?
>> Or where can I flee from your presence?
> If I ascend to heaven, you are there;
>> if I make my bed in Sheol, you are there.
> If I take the wings of the morning
>> and settle at the farthest limits of the sea,
> even there your hand shall lead me,
>> and your right hand shall hold me fast.

In the Wisdom of Solomon we are told that "the spirit of the Lord has
filled the world" and that "the one who holds all things together hears
every sound" (Wis 1:7). Then, later in the same book, we read:

> For you love all things that exist,
>> and detest none of the things you have made,
>> for you would not have made anything if you had hated it.
> How would anything have endured if you had not willed it?
>> Or how would anything not called forth by you have been preserved?
> You spare all things, for they are yours, O Lord,
>> you who love the living.
> For your immortal spirit is in all things (Wis 11:24–12:1).

The thought here is that things exist only because God loves them
and because the Spirit of God dwells in them. The indwelling Spirit is the
expression of divine love enabling creatures to exist and to evolve. The tradition of Trinitarian theology can be shown to support this line of thought.
In the dynamism of Trinitarian life, the Person of the Holy Spirit represents the ecstasy and excess of divine love. The Spirit is the *ek-static* one,
the one who goes out from the self to the other. Within the Trinitarian relations, because of the Spirit, the divine life of love cannot be thought of
as a self-sufficient love between two Persons. In the Trinity love breaks
beyond itself, beyond the love of the two, to involve the one whom

Richard of St. Victor called the "condilectus," the one who shares in love for the beloved and is loved with the beloved. The Holy Spirit makes it impossible for the love of the Trinity to be thought of as simply a face-to-face love of two Persons.

In Trinitarian thought, the Spirit is the ecstatic one who, in the divine choice to create, goes beyond the divine communion to what is not divine and brings what is not divine into relation with the divine Persons. As Christian Duquoc has said, the Holy Spirit makes the divine communion open to what is not divine. The Spirit is the indwelling of God where God is, in one sense, *outside* God's self.[16] The Spirit is God's *ecstasy* directed toward what is God's *other,* the creature. The Spirit as the excess, the ecstasy, of divine love brings creation into relationship with the divine life.

The Spirit expresses the dynamism and abundance of the divine communion. In the free divine choice to create a universe of creatures, it is the Spirit's role to be the dynamic and free overflow of divine communion that embraces the creatures. Yves Congar, in his three-volume work on the Holy Spirit, sees the Holy Spirit as absolute *Gift.*[17] He understands the Spirit as the Gift of divine communion *(koinonia)* directed toward creation. Congar points out that the Holy Spirit is always understood as a "going out," an "impulse," an "ecstasy." Following Duquoc, he, too, sees the Spirit as God "outside" God's self, God in creatures, God in us. He sums this up with the idea that God is Love and Grace, and says that "Love and Grace are hypostasized in the Spirit."[18] This means, he writes, that "the Holy Spirit, who is the term of the communication of the divine life *intra Deum,* is the principle of this communication of God outside himself and beyond himself."[19]

All this suggests that what Paul calls the *"koinonia* of the Holy Spirit" (2 Cor 13:13) can be understood to describe not only the Spirit's work in building the Christian community, and not only our adoption into the life of God (Rom 8:15-16), but also the Spirit's presence to all creation. The redemptive and liberating *koinonia* experienced in the Christian community was already understood by Paul as a pledge and foretaste of the reconciliation and communion of all things in Christ: "The creation itself will be set free from its bondage to decay and will obtain the freedom of the glory of the children of God" (Rom 8:21).

16. See Christian Duquoc, *Dieu différent* (Paris: Cerf, 1977) 121–122. See also Yves Congar, *I Believe in the Holy Spirit,* trans. David Smith (New York: Seabury Press, 1983) 3:148.

17. Congar, *I Believe in the Holy Spirit,* 3:144–154.

18. Ibid., 149.

19. Ibid., 150.

Jürgen Moltmann takes this insight further. He insists that the re-deeming Spirit of Christ and the creative life-giving Spirit of God are one and the same.[20] The *koinonia* of the Holy Spirit can be understood to embrace creation in all its complexity. Moltmann brings out the interrelation between the *koinonia* of the Spirit at work in the Church and the *koinonia* of all creation. He writes that the experience of the Holy Spirit in the community of the Church "leads of itself beyond the limits of the church to the rediscovery of the same Spirit in nature, in plants, in animals, and in the ecosystems of the earth."[21] The experience of the Spirit in the Church carries Christianity beyond itself to the greater community of all God's creatures. For Moltmann, then, "the community of creation, in which all created things exist with one another, for one another and in one another, is also the fellowship of the Holy Spirit."[22] He sees creation as "aligned toward community" and as "created in the form of communities." He sees it as appropriate to talk about a "community of creation" and to recognize the operation of the life-giving Spirit of God in the trend toward relationship in created things.[23] Furthermore, Moltmann suggests that this community of creation, which is inspired by the Holy Spirit, is meant to lead through evolutionary history, not to simple diversity or simple unity, but to "differentiated community that liberates the individual members belonging to it."[24]

In this section I have been proposing that it is faithful to the Christian tradition to understand the distinctive role of the Holy Spirit in ongoing creation in these terms: the Spirit is the one who *empowers* the evolutionary unfolding of creation precisely as the one who relates to each creature, bringing each into *communion* with the Trinity, and thus undergirding and enabling the communion of creatures among themselves. This is a differentiated communion, because each creature is loved and respected precisely for what it is and for its own precise participation in the ecological whole. The Creator Spirit is present in every flower, bird, and human being, in every quasar and in every atomic particle, closer to them than they are to themselves, enabling them to be and to become. Where the Spirit is, there also are the other Persons in the divine communion that is the being of God. The Spirit always unites creatures in Christ, in communion with the one who is the Source of all being. But it is the Creator Spirit who is the divine presence deep down in all things. It is this intimate

---

20. Jürgen Moltmann, *The Spirit of Life: A Universal Affirmation* (Minneapolis: Fortress Press, 1992) 10.
    21. Ibid.
    22. Ibid.
    23. Ibid., 225.
    24. Ibid., 228.

presence that enables creatures to exist and to transcend themselves in evolutionary change. The Spirit, present to every creature, brings every part of our universe into a creaturely communion with the triune God. The Spirit is the ecstatic gift of divine communion with creatures, whereby each creature exists and evolves in an interrelated world and whereby each creature is caught up with the dynamism of divine shared life.

A Proper Role for the Spirit of God in Creation

An obvious objection to any attempt to name what is distinctive about the role of the Spirit of God in creation is in terms of the traditional axiom that the Trinity's actions toward creation are one.[25] In the light of this axiom, is there any place for attempting to clarify a specific and proper role for the Spirit in creation? I will suggest that it is possible to hold that there is a proper role of the Holy Spirit in the one divine work of creation. This will, of necessity, involve some technical discussion, and in order to make this section as clear as possible, I will develop a line of thought with the aid of four summary proposals.

Proposal 1: *In creation and redemption, the Trinitarian Persons act only in profound communion and in undivided unity with one another; but this one undivided action does not exclude a* proper *role for each Person.*

The Greek theologians of the fourth century were convinced of the unity of divine action with regard to creation, but they did not see this unity as opposed to the distinction of Persons. Athanasius, for example, writes:

> There is, then, a Triad, holy and complete, confessed to be God in Father, Son and Holy Spirit, having nothing foreign or external mixed with it, not composed of one that creates and one that originated, but all creative; and it is consistent and in nature indivisible, and its activity is one. The Father does all things through the Word in the Holy Spirit.[26]

Athanasius goes on to specify that in the one act of creation, the first Person is the "beginning" and the "fountain" of creation, while the Word is the one "through" whom things are created, and the Spirit the one "in" whom things are made.[27] Gregory of Nyssa, too, insists that the Trinity acts as one, but according to the distinctive pattern of the divine relations:

---

25. This axiom finds expression in the Second Council of Constantinople (553) and at the Council of Florence (1442), as well as in the work of major theologians of the Eastern and Western traditions.

26. *Letters to Serapion*, 1, 28, in *The Letters of Saint Athanasius Concerning the Holy Spirit*, trans. C.R.B. Shapland (London: Epworth Press, 1951) 134–135.

27. Ibid.

"Every operation which extends from God to the creation, and is named according to our variable conceptions of it, has its origin from the Father, and proceeds through the Son, and is perfected in the Holy Spirit."[28]

In the West, Augustine's emphasis on the radical unity of the Trinity, and the unity of action with regard to creation, set the pattern of thought that found expression in the famous axiom that all the Trinity's actions *ad extra* are one. This axiom was taken up in Church teaching and, in my view, it is clearly right in what it affirms. God's actions with regard to creation are to be attributed to the one God. But what is often not attended to in the Western tradition is the role of the divine Persons in this one act of creation. With the emergence of a developed concept of appropriation in medieval Western theology, any real distinction of the Trinitarian Persons in the work of creation tended to be denied. The tendency was to think that anything that can be said distinctively of any one Person in relation to creation is really true of all; it is only *appropriated* to the specific Person and is not *proper* to that Person.

What does this theological concept of appropriation involve? Thomas Aquinas says that "to appropriate simply means to connect a thing that is common to something particular."[29] In Trinitarian terms, it means to speak of an action that is properly common to the whole Trinity as though it were the particular action of one Person. The idea is that when we say, for example, that all things are created by the power of the indwelling Spirit, we are taking what really belongs to the whole Trinity, to the one nature, and applying it to one particular Person. According to the theology of appropriation, it is legitimate to do this on the basis of a resemblance between the particular Trinitarian Person and the work, as long as it is acknowledged that it is *only an appropriation and not proper to the Person.*

I believe that there is an important place in theology for this understanding of appropriation. In the Christian Scriptures, in liturgy, in theological works, and in popular piety, we find attributes applied to one Person that properly belong to all three. Jesus is called, for example, "the resurrection and the life" in John (11:25), but other New Testament texts attribute resurrection to the Spirit and to the Father. The Creed attributes creation to the Father, although it is well understood that the whole Trinity is involved in creation.

What I want to suggest is that the theology of appropriation has an important and fundamental place at one level of theology and that it is

28. "On Not Three Gods," *Nicene and Post-Nicene Fathers,* 2nd series (Grand Rapids: Wm. B. Eerdmans, 1979) 5:334.

29. *De veritate,* q. 7, a. 3. See his *Summa Theologiae,* 1a, q. 37, a. 2, ad 3; q. 38, a.1, ad 4; q. 39, a. 7; q. 45, a. 6.

often misused at another level. Where it is used to caution against attributing a work such as creation *exclusively* to one Person, it has an indispensable role to play. There will always be the need to qualify the phrase "Creator Spirit," at least implicitly, with the idea that the other two Persons are also involved in creation. Creation cannot be attributed to one Person without reserve. But when the theology of appropriation is used at a second level, to deny that there is anything distinctive about, for example, the Spirit's work in creation, then I believe that it is being misused. I will try to show that there are things that can be said about the Spirit's role in creation that are proper to the Spirit.

Proposal 2: *A foundation for a theology of the proper role of the Spirit in creation can be found in the work of contemporary theologians who discuss the proper roles of the Trinitarian Persons in the incarnation and in the Pentecost event.*

Many contemporary theologians are dissatisfied with an unqualified use of the theology of appropriation, above all when it is used to deny the real and proper roles of the divine Persons in the incarnation and our sanctification. Karl Rahner, for example, has argued convincingly that the incarnation is clearly a case where a divine presence in the world is not merely *appropriated* to one Person but is *proper to the Person of the Word.* He does not deny that the whole Trinity is causally involved in the incarnation but insists that in the incarnation of the Word something occurs that is proper to one Person and can be predicated of that one Person alone.[30] Rahner rejects as misleading and false the idea that any other Person of the Trinity might have become incarnate.[31] In a similar way, theologians such as Heribert Mühlen and David Coffey have pointed out that the outpouring of the Spirit at Pentecost and in the work of sanctification must be understood as proper to the Holy Spirit.[32]

In a different but not opposed approach to this question, Karl Rahner and Yves Congar have both argued that in the work of sanctification we have a real and proper relation to *each* of the divine Persons.[33] What is at stake here is a fundamental issue of spirituality. Do we relate to the divine Persons or simply to the one divine nature? Rahner and Congar insist that

---

30. Karl Rahner, *The Trinity* (New York: Seabury Press, 1974) 23.

31. Ibid., 28–30.

32. Heribert Mühlen, *Der Heilige Geist als Person* (Münster: Aschendorff, 1963); David Coffey, "A Proper Mission of the Holy Spirit," *Theological Studies* 47 (1986) 227–250; *Grace: The Gift of the Holy Spirit* (Sydney: Faith and Culture, 1979).

33. Rahner, *The Trinity,* 34–38; "Some Implications of the Scholastic Concept of Uncreated Grace," *Theological Investigations* (London: Darton, Longman and Todd, 1961) 1:319–346; Congar, *I Believe in the Holy Spirit,* 2:85–92.

we relate to the Persons in a proper and distinctive way. They understand the relation between the divine Persons and graced humanity in terms of formal causality. By this they mean to indicate a causality that involves a real *self-communication* of God to human beings in the Word and in the Spirit. This divine self-communication really transforms human beings constitutively, without compromising divine transcendence or holiness.[34] Human beings are really assimilated to Jesus Christ through the action of the Spirit and become themselves adopted daughters and sons of God. According to Congar and Rahner, this assimilation involves real and proper relations to the divine Persons.

At first glance there can appear to be a conflict between the claim that sanctification is a work that properly belongs to the Spirit and the claim that in our sanctification we relate in proper and distinct ways with each of the divine Persons. But this is only an apparent conflict. There is no contradiction in holding that our sanctification is due to the bestowal of the Spirit, on the one hand, and that this bestowal of the Spirit brings us into a real relation with the whole Trinity, on the other. In fact, I would argue that this is the most obvious interpretation of important biblical texts such as Romans 8:14-17.

Furthermore, I would suggest that when contemporary theology rightly claims a proper role for the Word of God in the incarnation and a proper role of the Spirit in Pentecost, this does not exclude, but necessarily includes, a distinctive and proper role for the other Persons in these same events. If one adopts a New Testament Spirit Christology, one must say that not only the Word but also the Spirit has a proper role in the incarnation. It is God's bestowal of the Spirit that constitutes Jesus of Nazareth as messianic Son of God. On the other hand, it seems clear that not only the Spirit but also the Word has a proper role at Pentecost. The outpouring of the Spirit is also a Christ-event. It constitutes the Church as the Body of Christ, so that the Church is "in" Christ and the Church is Christ.

Both great events—the incarnation of the Word and the outpouring of the Spirit—involve not only the Word and the Spirit but also the divine Person who is Unoriginate Origin and Source of all. So while it is true that the incarnation is distinctively and properly the mission of the Word of God, it also involves the *proper* roles of the other Trinitarian Persons. And while Pentecost (and our sanctification) involves the distinctive and proper mission of the Spirit, it also involves the *proper* participation of the other Persons.

---

34. In order to show that God can constitutively transform human existence by self-communication, *without thereby compromising divine transcendence,* Rahner and Congar often speak of a "*quasi*-formal causality."

This same line of thought can open out into a theology of creation in which each divine Person can be understood to have a proper role. This is not to collapse the distinction between creation and the economy of grace, but to point to the inner relationship between God's self-expression in creation and God's self-communication to us in salvation history. An initial argument for a proper role of the Spirit and the Word of God in creation is that this is the most obvious meaning of a number of biblical texts. Texts such as Genesis 1:1-2 and 2:7, for example, point toward the concept of the Spirit as the Life-giver, while texts like John 1:1-5 and Colossians 1:15-20 suggest an exemplary theology of Jesus Christ as the Word and Wisdom through and in whom all things are created.

Proposal 3: *A first argument for a proper role of the Spirit in creation is that ongoing creation is best understood as a dynamic relationship between each creature and the Trinity; such a relationship approach would involve the distinct Trinitarian Persons.*

The kind of causality involved in creation is precisely a relational causality. The simple model of efficient causality is not an adequate way of understanding creation by a Trinitarian God. Karl Rahner is right to insist that creatureliness is not a particular instance of a causal relationship that can be found elsewhere among creatures.[35] We cannot fit God's creation into a system of cause and effect that we find operating between one empirical reality and another. The relation of Creator and creature is absolutely unique. There is an infinite difference between divine "causality" in ongoing creation and the kinds of causality we find at work in the world.

What can be said about the kind of causality that the Trinitarian God exercises in creation? If God's being is radically relational, if God's being *is* communion, then this suggests something about the whole of reality. It suggests an ontology of relation. A radically relational God might be thought to create a world that is relational to its core. When we turn to contemporary science, we find that cosmology, quantum physics, evolutionary biology, and the whole range of ecological sciences suggest a relational worldview. Theology and science can meet in a view of reality as relational—in a relational ontology.

In a relational ontology, God, understood as Persons-in-communion, can be thought of as creating a relational universe precisely in and through the relation of ongoing creation. In the relational causality of creation, the divine Persons would be understood as giving expression to *themselves* in what is not divine. It is this dynamic divine self-expression that constitutes creatures and allows them to be their own creaturely selves. It gives them

35. Rahner, *Foundations of Christian Faith*, 76–77.

existence. It empowers the universe for becoming. It is not yet the hypostatic union or God's self-communication in the Spirit, but it is directed toward these. It is precisely the self-giving of God in creation and nothing more. But it *is* a self-giving.

If God is Persons-in-communion, and if ongoing creation is a relational act whereby each creature is enabled to be and to become in a community of creatures, then this suggests that each divine Person is to be understood as distinctly and properly engaged in the one act of creating. If creation is, from the side of God, a personal self-giving, a personal relationship, this would seem to involve what is distinctive and proper to the divine Persons.

Proposal 4: *A second argument for a proper role of the Trinitarian Persons in creation is that what are distinctive about the Trinitarian Persons (their relations of origin) come into play in the one work of divine creation.*

Creation is a being-caught-up-in-relation with the divine Persons-in-communion. I want to suggest that this involves each of the divine Persons precisely in what distinguishes them from one another: the first Person as Source, the second as eternal Word generated in divine self-expression from the Source of all, the third as the Spirit who is ecstatically breathed forth from that same Source in the excess of divine Love.

First, I have argued above that it is faithful to the Christian tradition to see the Spirit as the power of becoming and the ecstatic gift of communion in creation. Second, the Word and Wisdom of God can be understood as the exemplar for all things, the "icon of God" in whom "all things" were created (Col 1:15). As Bonaventure has said: "Every creature is of its very nature a likeness and resemblance to eternal wisdom."[36] The Word of God, who expresses the self-distinction in God, mediates in creation (Heb 1:2; John 1:3) as the basis of the self-distinction and independent existence of all creatures.[37] And, third, every creature springs ultimately from the Source of all being, the one whom Bonaventure called the *fontalis plenitudo* (the "Fountain Fullness"). Creation is a relation by which the ecstatic Spirit is given to creatures empowering existence and

---

36. *Omnis enim creatura ex natura est illius aeternae sapientiae quaedam effigies et simultudo* (*Itinerarium*, 2.12). In the *Hexaemeron*, Bonventure writes: *Unde creatura non est nisi quoddam simulacrum sapientiae Dei, et quoddam sculptile* (12).

37. Pannenberg writes that the Word acts in creation "as the principle not merely of the distinction of the creatures but also of their interrelation in the order of creation." He shows how in the incarnation, the working of the Logos in all of creation, as the principle of self-distinction, comes to unthinkable fulfillment. This, too, occurs through action of the Spirit: "The Spirit mediates the working of the Logos in creation as also in the incarnation." Wolfhart Pannenberg, *Systematic Theology* (Grand Rapids, Mich.: Wm. B. Eerdmans, 1994) 2:32.

evolution, bestowing on them their own appropriate communion with divine Wisdom, the one in whom all things are created, and with the one who is the Source and Fount of all goodness and life.

As I have suggested above, in the Trinity the Spirit can be understood as expressing the *ek-stasis* of the divine communion. It is thus "in the Spirit" that the divine Persons-in-communion reach out to what is not God. The Spirit continues in the economy of creation and salvation what the Spirit always is in the divine life. I find this idea expressed in an official catechetical text on the Holy Spirit prepared by the Roman Catholic Church in preparation for the Holy Year 2000. It states that God places Godself in communion with us "in the Spirit," and it explains that "this 'in the Spirit' fills the infinite distance which separates the uncreated from the created, God from human beings, and becomes God-for-us, God-with-us, and God-among-us."[38] In this text the Spirit is seen as the "final touch" through which God unites with creatures, "saves them from non-existence, sustains them, renews them, and directs them toward their fulfillment."[39]

The Spirit is the touch of God that unites creatures in communion with the Trinitarian God, saving them from non-existence, sustaining, renewing, and directing them toward their fulfillment precisely as the ecstatic gift of divine communion with all that exists. Because this is directly related to what is distinctive about the Spirit's inner divine life, I think that there is every reason to conclude that this is a distinctive and proper role for the Spirit in the work of creation. The Spirit is the ecstatic gift of communion only in profound unity with the other divine Persons. Nevertheless, this work of *empowering* and *communion-making* between creatures and the Trinitarian God can be understood as a *proper* role of the Holy Spirit. It is not simply that this work is appropriated to the Holy Spirit; it can be thought of as proper because it is a reflection in the economy of what distinguishes the Holy Spirit within the Trinitarian relations.

The Spirit of God as the Ecological Spirit

Ecology is concerned with the interrelated systems that support life on our planet. I have been proposing in this chapter that at the theological level this can be understood as very much the domain of the Spirit of God.

---

38. *The Holy Spirit, Lord and Giver of Life,* prepared by the Theological-Historical Commission of the Great Jubilee of the Year 2000, trans. Agostino Bono (New York: Crossroad, 1997) 22.

39. Ibid., 31. This document argues strongly for the distinctiveness of the Trinitarian roles in creation and salvation, but then, in one place at least (p. 19), says that this is by appropriation.

The Creator Spirit's role is to enable each creature to be and to become, bringing each into relationship with other creatures in both local and global ecological systems, and in this process of ongoing creation, relating each creature in communion within the life of the divine Persons-in-communion.

This means that forests, rivers, insects, and birds exist and have value in their own right. They are not simply there for human use. They have their own integrity. They exist as an interdependent network of relationships in which each creature is sustained and held by triune love. They manifest the presence of the Spirit as the ecstasy and fecundity of divine love.

The Christian tradition, then, in its theology of the Spirit, provides a theological foundation that can give expression to the experience discussed at the beginning of this chapter, namely, the experience of a particular place, and of a particular ecology, as a place of wonder and mystery. This theology provides a foundation for seeing the experience of healing and of coming to wholeness that can occur in such places, not simply as encounters with the mountains or the sea and the creatures that inhabit them but also as an encounter with the Spirit who dwells in that place, bringing its varied creatures, including the human ones, into a life-giving relation with the divine communion. The arguments advanced here suggest that the Spirit is present in all creatures and in all places.

There is a further, urgent question that I am unable to address here. Why, in some circumstances involving, for example, animals playing with helpless prey or humans being killed by an earthquake, is it all but impossible to see nature as the place of the Spirit? Lorna Hallahan's essay in Chapter 6 of this volume addresses this kind of question directly. I attempt a response to it in other places.[40] Here I will say only that the Spirit, who is always present to all things, is to be understood not only as rejoicing in creation but also as *suffering* with suffering creation, "groaning in labor pains" (Rom 8:22) with creation until all is transformed in Christ.

I have been suggesting that an important foundation for ecological theology is the conviction that the Spirit of God is creatively and lovingly present to all creatures, and present in the whole of our interconnected planetary life. The earth, then, has a sacramental character—it symbolizes the divine that is present in it. Long ago, Ambrose of Milan saw the Spirit

40. As part of the same research project, I am following up these issues in a second article on the Holy Spirit and creation titled "Ecology and the Holy Spirit: The 'Already' and the 'Not-Yet' of the Spirit in Creation," *Pacifica* 13 (June 2000) 142–159. I have already developed some of these ideas in *The God of Evolution: A Trinitarian Theology* (New York: Paulist Press, 1999) 35–55. In this volume, see the essays by Gregory Brett, C.M. (Chapter 9) and Anthony Lowes (Chapter 7), which take up the issue of eschatology in relation to creation.

sweeping over the waters of creation not only as the Life-giver but also as the one who brings beauty to creation. Creation's resplendent beauty is a gift received from the Spirit of God.[41] Basil of Caesarea developed what might be called the beginning of a Christian ecological attitude to creation when he invited his hearers into a way of seeing all creatures: "I want to awake in you a deep admiration for creation, until you in every place, contemplating plants and flowers, are overcome by a living remembrance of the Creator."[42] The Holy Spirit is the immanent presence of the divine Trinity, the source of life and beauty, the one who groans and suffers with creation until it is brought to its consummation.

If Jesus of Nazareth can be understood as the human face of God in our midst, the Spirit can be thought of as God present in countless ways that are far beyond the limits of the human. If, in Jesus, God is revealed in specific human historical shape, in the Spirit, God is given to us in a personal presence that exceeds all human limits. In a lovely phrase used by Moltmann, the Spirit is the "unspeakable closeness of God," the one we encounter in the experience of deep connection with a place, in delight in trees, flowers, birds, and animals.[43] The Spirit is the closeness of God we experience in silence before the mystery of the universe. This is the same Spirit we experience at the heart of mutual friendship, in finding that there is a holy presence with us in times of suffering and grief, and even in what seems at first like nothing but absence and abandonment. It is this same Spirit who stirs within us in the experience of faith in the gospel, in participation in the eucharistic communion of the Church, in the sense of global solidarity with all human beings and with all God's creatures, and in our union with Jesus before the One he called *Abba* and who can also be called our beloved Mother. The Holy Spirit is inexpressible personal closeness in all these ways and in many others.

The presence of the Spirit cannot be limited by human expectations. The Spirit is radically beyond us: "The wind blows where it chooses, and you hear the sound of it, but you do not know where it comes from or where it is going" (John 3:8). The Spirit transcends our humanity and its preoccupations with itself, embracing all God's creatures. This is why the great symbols of the biblical tradition speak of the Spirit more effectively than anthropomorphic language or abstract theology. The Spirit is living water, the untamed wind, blazing fire. The Spirit pervades the whole universe and sees to "the depths of God" (1 Cor 2:10). To be in communion with this Spirit is to be in communion with the whole of creation.

41. Ambrose of Milan, *On the Holy Spirit*, II, 32.
42. Basil of Caesarea, *Homilies on the Hexaemeron*, VI, 1.
43. Moltmann, *The Spirit of Life*, 12.

All this is suggesting that Earth reveals. It is the place of encounter with the Holy Spirit. But this theology also commits us to repentance for our abuse of our planet, to the way of Earth healing. As Edward Echlin says, the injustice we have committed against the Earth community demands restitution of us. It requires literally that we roll up some of the asphalt that has been crushing out life. He writes that "ecological, holistic healing of God's Earth community is going back—to biodiversity, flourishing habitats in town and country, clear air and water, and closeness of people to nature."[44] This demands political action and long-term commitment. At the core of this is learning again to love the diversity of earth's creatures.

44. Edward Echlin, *Earth Spirituality: Jesus at the Centre* (New Alresford, U.K.: John Hunt Publishing, 1999) 129.

Barra and Turtle

Barramundi
And short necked turtle
Move free and easy
And the water—it's theirs.

You can catch them
With a fishing line or spear,
But only take what you need
So there's more next time.

They're very good eating,
Especially when barbecued in the coals
And shared with everyone . . .
We enjoy the togetherness.

Christine Rowan

Chapter 4

Enfleshing the Human

An Earth-Revealing, Earth-Healing Christology

Duncan Reid

> We believe in one Lord, Jesus Christ,
> the only Son of God,
> eternally begotten of the Father,
> God from God, Light from Light,
> true God from true God,
> begotten, not made, of one Being with the Father;
> through him all things were made.
> For us and for our salvation
> he came down from heaven,
> was incarnate of the Holy Spirit and the virgin Mary
> and became truly human.
> For our sake he was crucified under Pontius Pilate;
> he suffered death and was buried.[1]

As worshipers, many Christians recite these words every week. But do we consider their meanings? The central Christological affirmation of the Nicene Creed occurs in the two phrases "became flesh" *(incarnatus est)* and "was made human" *(homo factus est)*.[2] The first refers, clearly, to the Johannine prologue with its focus on the eternal Word of God which in Christ became flesh (John 1:14). The second amplifies and focuses the

---

1. "The Nicene-Constantinopolitan Creed," *Confessing One Faith: Towards an Ecumenical Explication of the Apostotlic Faith as Expressed in the Nicene-Constantinopolitan Creed (381),* Faith and Order Paper No. 140 (Geneva: World Council of Churches, Commission on Faith and Order, 1987) 7.

2. Greek: *kai sarkothenta/kai enanthropesanta.*

first by pointing to the story of Jesus as a particular human being. These two credal phrases are separated by the explanatory comment "of the Holy Spirit and the virgin Mary." This explanatory comment attempts to answer the question of how this incarnation, this becoming human, has occurred.

The argument of this chapter is very simple, namely, that the theological separation of these two key phrases, the affirmation of the "becoming flesh" and its elucidation in terms of "being made human," needs to be overcome. My contribution to this volume will attempt a task of theological retrieval by seeking to reconnect these two terms so that the fleshliness of Christ's humanity is not overlooked. In doing this, I acknowledge that I will be drawing on a specific Christological tradition of an Alexandrian tendency, one that emphasizes Christ's flesh rather than the more Antiochene emphasis on Christ's humanity.[3] I also acknowledge that, contrary to the admirable current trend to write Christology *from below,* this will be an unapologetic Christology *from above.* This methodology may itself be seen as an exercise in theological retrieval, in this case seeking ecological themes where we might not have expected to find them, in a more traditional way of doing theology.

In the centuries after the formulation of the Creed, the dominant theological tradition widened the gap between the flesh and the human, so that the human could be thought of in an "unfleshed" way and the flesh could be subordinated to the human.[4] The essential biblical insight into human inclusion in the concept "all flesh,"[5] that is, all living things, could be forgotten in favor of an essentially disembodied notion of humanity or human

---

3. I use the term "retrieval" in much the way it has been developed in feminist theology and is used in this volume by Patricia Fox (Chapter 5) and James McEvoy (Chapter 11), that is, in the sense of rediscovering in the theological tradition a perspective (perhaps long neglected) of practical value. In the history of doctrine, Alexandrian Christology tends to emphasize the unity of the Word and the flesh in Jesus Christ, while Antiochene Christology is more interested in the full humanity and divinity of Jesus.

4. Paul Collins sees even such thinkers as Pierre Teilhard de Chardin and Thomas Berry, whom we might otherwise expect to help build an ecological Christology, continuing to drive this damaging wedge between the cosmic Christ and Jesus of Nazareth (Paul Collins, *God's Earth: Religion as If It Really Mattered* [Dublin: Gill and Macmillan, 1995] 235). On pp. 242–244 he emphasizes, as I do here, the Christological importance of the flesh.

5. Heb. *kol basar,* Gk. *pasa sarx* (e.g., John 17:2). S.v. *basar* in *Theological Dictionary of the Old Testament,* rev. ed. (Grand Rapids, Mich.: Wm. B. Eerdmans, 1977) 317–332, especially 319. The term *kol basar* ("all flesh") occurs in this sense in Genesis 6:17; 9:11; 9:15ff.; Num 18:15; Ps 136:25; Dan 4:9; Sir 40:8. S.v. *sarx* in *Theological Dictionary of the New Testament* (Grand Rapids: Wm. B. Eerdmans, 1971) especially 138–140. In passages like Genesis 6:12, "all flesh" is clearly inclusive of human beings.

nature. It then becomes the human ideal to rise above the flesh, and the resulting worldview is both anthropocentric and flesh-denying.[6]

Christology has continued to emphasize the humanity of Christ, Christ's relationship with us human beings, rather than the fleshliness or the embodiedness of Christ. This tendency obscures Christ's prior relationship with all flesh and indeed with all created things. The majority of even recent contributions to Christology focus on the humanity of Jesus rather than on his fleshliness and consequent kinship to all flesh.[7] The result is that books on Christology tend to avoid ecological themes, and books on ecotheology tend to avoid Christology.[8] This contrasts with a number of recent books on the doctrine of God, and the Holy Spirit, where an ecological perspective does indeed come to the fore.[9]

6. See Paul Santmire's "metaphor of ascent," which understands religious experience in terms of encountering transcendence as a mountaintop event "far removed from the mundane things of the earth below," and in a feeling of being "carried away from every earthly constraint" (H. Paul Santmire, *The Travail of Nature* [Philadelphia: Fortress, 1985] 18; 182ff.).

7. This is perhaps an unfortunate consequence of doing Christology from below. But Christ has a prior relationship to all flesh, and indeed to all creation, insofar as he is the eternal Logos who in creation orders all things. This relationship is not dependent upon the incarnation. The Johannine prologue is important here as a corrective to any suggestion that a relationship as such is established only with the incarnation. But it is proper to the eternal Logos that it become incarnate, a body subject to death, thus transforming the hierarchical creator-creation relationship into a more intimate kinship relationship.

8. In opening his discussion of the cosmic Christ, Jürgen Moltmann remarks that despite the presence of this theme in theology at least since 1951, and its prominence at the 1961 New Delhi WCC Assembly, it fails to rate a mention in two of the major contributions to Christology of the past thirty years (Jürgen Moltmann, *The Way of Jesus Christ: Christology in Messianic Dimensions* [London: SCM, 1990] 374, n. 3). The authors referred to are Pannenberg and Kasper, but without too much trouble we could expand Moltmann's list. Paul Santmire senses this uncomfortable relationship between Christology and ecology when he points out that what is often described as the "nature mysticism" of Francis of Assisi is in fact a Christ-mysticism, one that emphasizes Christ's descent, or *kenosis*. What is significant is that this ecologically oriented Christ-mysticism is so often not recognized. See H. Paul Santmire, *The Travail of Nature*, 109–113. In this section (112) Santmire helpfully offers an alternative translation of Francis's Canticle of the Sun, claiming that the author intended the creatures to be regarded as holding an intrinsic value rather than the merely instrumental value suggested by the more common translations.

9. On the doctrine of God, e.g.: Catherine LaCugna, *God for Us: The Trinity and Christian Life* (Edinburgh: T&T Clark, 1991); see also the chapters in this volume by Patricia Fox (5), Lucy Larkin (8), and Anthony Lowes (7). On the Holy Spirit, e.g.: Jürgen Moltmann, *The Spirit of Life: A Universal Affirmation* (London: SCM, 1992); Peter C. Hodgson, *Winds of the Spirit: A Constructive Christian Theology* (Louisville: John Knox Press, 1994); Geiko Müller-Fahrenholz, *God's Spirit: Transforming a World in Crisis* (New York: Continuum; Geneva: World Council of Churches, 1995); Mark Wallace, *Fragments of the Spirit: Nature, Violence and the Renewal of Creation* (New York: Continuum, 1996). See also Chapter 3 by Denis Edwards in this volume. The reasons why the doctrines of God

Perhaps surprisingly for those who would see John's Gospel as verging on gnosticism, it is the Johannine prologue, together with the oblique reference to it in the Creed, that keeps alive the biblical insight that the human being is integrally of the earth, a part of all flesh. It is this insight that opens possibilities of an ecological Christology. It is the story of Jesus as a human being that, when separated from this insight of the Johannine prologue, endangers this possibility by driving Christology irreversibly in the direction of anthropology. In the course of this chapter I want to test this argument by looking at the theological tradition and to propose two guiding principles.[10] The first of these will be a specifically Christological principle for developing an earth-friendly Christology, and the second a derivative principle of a more general, anthropological nature. These principles, which can be drawn from the wording of the Creed, are: that *Christology will hold together the affirmation that Christ has come as a human being with the prior affirmation that the second Person of the Trinity has become flesh*; and that *specific human concerns will be considered only in the wider context of the concerns of all flesh*. But first we need to deal briefly with some potential problems.

Flesh, Body, Matter

Several problems are foreseeable. The term "flesh" is not always used in the Bible in the inclusive way I have suggested. In 1 Corinthians 15:39, for example, each species is understood as having its own particular flesh. The argument to be developed here depends on an appeal to the older and more comprehensive tradition expressed in the term "all flesh" *(kol basar)*. "Body" is not always equivalent to "flesh," and again the passage from 1 Corinthians 15 illustrates this. The heavenly body referred to in verse 40 is contrasted to the earthly body, and this is sometimes understood as a contrast between a material body and something immaterial. There are two points to be made here. First, it would be a mistake to consider heaven and the heavenly as independent of creation. Heaven, in the Bible, is a part

---

and the Spirit might lead to ecotheology are not hard to discern. The doctrine of God is increasingly likely to be developed in Trinitarian terms, where a perichoretic notion of the divine Persons in mutual embrace has clear analogies to the ecological vision of a web of life in the natural world. The Holy Spirit is named in the Creed the "Life-giver" *(zoopoion/vivificantem)*, a title suggestive of deep involvement in the life processes of the created order.

10. By "guiding principles" I mean something akin to the rules or regulative statements that both Dietrich Ritschl (*The Logic of Theology: A Brief Account of the Relationship Between Basic Concepts in Theology* [London: SCM, 1986]) and Kathryn Tanner (*God and Creation in Christian Theology: Tyranny or Empowerment?* [Oxford: Basil Blackwell, 1988]) see as guiding theological discourse.

of created reality and may have more in common with matter than is often assumed.[11] Second, I would appeal, once again, to the close relationship between the terms "body" and "flesh" attested to elsewhere in the biblical and extrabiblical traditions.[12]

A further objection could be that flesh is not inclusive of matter in the way that matter is inclusive of flesh, so that even to establish a kinship between the incarnate Christ and non-human animals may still be without ecological significance—that is, significance for ecosystems, with the plant life and inanimate things that constitute them. To answer this, it may be sufficient to appeal to the contemporary ecological worldview itself. Animals, like human beings, are simply inconceivable without the environments in which they live and which support them. But we need go no further than the biblical and theologial traditions, and again here my argument depends on making connections rather than distinctions. The second creation story, beginning at Genesis 2:4b, makes a clear link between humanity and the earth, a link that finds echo in many other biblical passages. The link is brought out playfully yet profoundly in the verbal connections in Genesis 2:7 between the human being *(ha adam)* and the dust of the ground *(ha adamah)*. Human life is clearly connected with and contingent upon inanimate matter in this story. A further link between these terms *(ha adam, ha adamah)* and blood *(dam* or *dammim)* is made explicit in Genesis 4:10 when the blood of Abel cries out from the ground. Blood, for the biblical writers, is a common element in all animate things.[13] Just as the significance of blood in New Testament soteriology can now be recognized as incomplete without feminist insights into its symbolism, perhaps it is also incomplete without taking into account this enduring theme of the commonality of blood in all animals.[14]

These problems can, I believe, be laid to rest for the purpose of the present argument.[15] There is, however, a bigger issue that emerges from

11. Jürgen Moltmann, *The Coming of God: Christian Eschatology* (Minneapolis: Fortress, 1996) 303; M. Welker, *God the Spirit* (Minneapolis: Fortress, 1994) 137ff.

12. See *soma* in *Theological Dictionary of the New Testament,* 1024–1094. There is a close relationship between *sarx* and *soma* in Philo (1051), the New Testament (1075), and the apologists (1083). The Hebrew *basar* is on occasions translated as *soma* in LXX (1044). The significance of body for ecotheology is explored in this volume in Chapter 6 by Lorna Hallahan and in Chapter 10 by Phillip Tolliday.

13. E.g., Lev 17:14; Deut 12:23. Cf. Acts 15:20.

14. E.g., Janet Gaden, in John Gaden, *A Vision of Wholeness* (Sydney: E. J. Dwyer, 1994) 245ff.

15. These issues have been discussed in depth by some of the most significant theological minds. Karl Barth (*Church Dogmatics* [Edinburgh: T&T Clark, 1932ff] vol. II/2, 51ff.) locates his doctrine of the election of a particular people within the larger framework of the election of all creation: "He does the general for the sake of the particular. Or to put

these questions, namely, a deep-seated ambivalence in the theological tradition to flesh. This ambivalence can be explained in terms of fears of sexuality and of that which is other, fears also of pain and mortality. There is a similar deeply entrenched and perhaps entirely understandable sense that our humanity lies beyond our fleshliness or embodiedness. This second phenomenon is not quite so easily dismissed, because it is grounded in our personal sense that those we love, especially in the face of pain or dying, are and always will be somehow more than the pain or disability they may at present be suffering. This is clearly too large a subject to embark upon here.[16] What it means, though, is that this present discussion is nothing more than a beginning, a pointer to a way forward.

Two Samples from the Tradition

To chart the gap between the flesh (or embodiedness) of Christ and Christ's humanity, I want to drill into the ancient lakebed of the tradition and examine, as it were, two core samples, Athanasius and Anselm. This choice of samples should not be too surprising: Athanasius's treatment of Christology is known by the title *On the Incarnation of the Word*.[17] Anselm's best known treatment of the same topic is worked out under the heading *Cur Deus Homo? (Why Did God Become Human?)*.[18] These two samples should be taken, not so much as indicative of a progess or regress within the tradition, but as examples of two tendencies or themes that have emerged at different times and places.

---

it in another way, He does the general through the particular, and in it and with it" (53). Karl Rahner (*Foundations of Christian Faith: An Introduction to the Idea of Christianity* [London: Darton, Longman & Todd, 1978] 178ff.) begins his Christology within the presupposition of an evolutionary view of the world involving "the unity of all created things," in which "the higher order always embraces the lower order as contained within it" (186). John Zizioulas ("Preserving God's Creation: Three Lectures on Theology and Ecology," *King's Theological Review*, 12 [1989] 1–5), in common with the Eastern Orthodox tradition in general, insists that in the lifting up *(anaphora)* of bread and wine and other material things in worship, the whole material world is presented before God, and that the biblical image of God in humanity must be understood in bodily terms.

16. See Chapter 6 by Lorna Hallahan in this volume.

17. *Oratio de incarnatione Verbi*, in J.-P. Migne, *Patrologia Graeca* (PG) 25, 95–198. Translated by Archibald Robertson, in E. R. Hardy, ed., *Christology of the Later Fathers* (Philadelphia: Westminster Press, 1964).

18. *Cur Deus Homo?* in J.-P. Migne, *Patrologia Latina* (PL) 158, 359–452. Translated by Joseph M. Colleran, in Anselm of Canterbury, *Why God Became Man and The Virgin Conception and Original Sin* (Albany, N.Y.: Magi Books, 1969).

Athanasius (c. 296–373)

Athanasius, the great fourth-century bishop of Alexandria, had attended the Council of Nicaea as a young man and went on to become a lifelong defender, at considerable personal cost, of the Christology articulated there. He opens his work with a reference to the Word's becoming human, but then follows an explanation that speaks of the Savior "wearing a body" and the Word being "manifested in a human body for our salvation."[19] Soteriology is placed in a prominent position here, and Athanasius repeatedly emphasizes the embodiedness of Christ's appearing.[20] In chapter 2 he alludes to God's creation of matter as contingent and not co-eternal with God. The subtext is that matter stands in a particular relationship to God, such that it will be important that the Word become matter. Human beings are matter first and foremost, and only then distinguished from other creatures by virtue of having "as it were a kind of reflection of the Word."[21] This likeness is emphasized by a play on *logos* ("word") and *logikos* ("rational"). This correspondence has been corrupted by human wrongdoing, but God as Creator is unwilling to see God's "own work to be ruined."[22] It is not just the human creature that is ruined, but all creation suffers.

Athanasius goes on to insist that "no part of creation is left void" of the presence of the Word, which fills all things while remaining also present with the Father.[23] Even so, the Word chooses to take "a body of our kind," that is, "a body capable of death."[24] This "body of our kind" does not, then, have to be read exclusively as a human body but can be understood inclusively as a body that, like all bodies, is subject to death. This more general category of bodies is only then narrowed down to the human body in which the Word appeared, because it is through human activities that the corruption of creation has come about.[25] The Word needed to become flesh, but then also specifically human flesh, not because humans are somehow better than all other flesh, but on the contrary, because humans are responsible for the corruption and suffering of the rest of creation.

---

19. Ch. 1 (PG 25, 97).
20. E.g., ch. 4 (PG 25, 104).
21. Ch. 3 (PG 25, 101).
22. Ch. 5 (PG 25, 105); Ch. 6 (PG 25, 105).
23. Ch. 8 (PG 25, 109). This theme is reiterated in chapter 17, where Athanasius argues that the Word is not "circumscribed in the body" but is active throughout all creation, "distinct in being from the universe, but present in all things by his own power." So during the incarnation the Word continues active in three ways: "walking as human, as the Word quickening all things, and as the Son . . . dwelling with his Father" (PG 25, 125).
24. Ch. 8 (PG 25, 109); Ch. 9 (PG 25, 112).
25. Ch. 10 (PG 25, 113), where Athanasius cites 1 Corinthians 15:21-22 in support.

"This then is the first cause of the Savior's becoming human," concludes Athanasius, who continues throughout to hold together these two elements of becoming flesh and becoming human.[26]

Athanasius is not free of Santmire's metaphor of ascent but uses it to suit his purpose.[27] Because human beings look down "seeking about for God in nature and the world of sense," the Word chooses to meet human beings in a body, thus "meeting the senses of all people halfway."[28] For this reason it was not enough for the Word simply to become flesh, subject to death, for this would have passed unnoticed;[29] it was also necessary for the Word to appear through certain works, showing by signs who he was for this salvation to be effective.[30] Thus the humanity of Christ is necessary, but in a way that is logically secondary to the Word having become flesh.

Chapters 20–32 focus on the significance of the crucifixion. The debt of all people could be paid by this one death in the body, a public, shameful death.[31] That this is a death for all people is signified by the spreading of Christ's hands. Further, it is a death, signified by Christ's being lifted up, to "clear the air of the malignity of the devil and all demons."[32] Athanasius sees this as a cosmic cleansing that spreads over and purifies all of polluted nature. Constantly he repeats that this death overcomes the death that afflicts not just human bodies but all bodies, the Savior having "appeared in the body" and taken "a body for the salvation of all."[33] In context, it is not too fanciful to read this as pointing to the salvation not just of all human bodies but of all bodies.

This matter of Christ's fleshly embodiment is explored further when Athanasius turns to address particular objections. Where the Jews object to the idea that the Word of God might become human, the Greeks, significantly, object to the idea that the Word of God might be made manifest in a body.[34] The Greeks rightly understand the universe as a great body, says Athanasius. But if the Word of God is in the universe, which is a

26. Ch. 10 (PG 25, 113); e.g., in ch. 14: "he sojourns here as human, taking to himself a body like the others, and from things of the earth" (PG 25, 121).

27. See n. 6 above.

28. Ch.15 (PG 25, 121–123).

29. Ch. 16 (PG 25, 123–125).

30. Ch. 18 (PG 25, 127).

31. The short recension includes "For the body suffered and died in the nature of bodies" (ch. 24, in *Christology of the Later Fathers*, 78–79).

32. Ch. 25 (PG 25, 140).

33. Ch. 29 (PG 25, 145); Ch. 32 (PG 25, 152).

34. Ch. 33 (PG 25, 177); Ch. 41 (PG 25, 167-9), see also ch. 43. Could our theological hesitations about a fully embodied Christology perhaps be (PG 25, 172–173) a lingering effect of our European bondage to Greek philosophy?

body, why is it absurd to say this Word cannot unite itself with a particular human body, especially as humankind is a part of this whole? If it is unseemly for the Word to unite with a human body, then it is surely absurd to claim, as Greeks do, that the Word can be manifest through the whole!

Only after this discussion does Athanasius turn to the question as to why the Savior should appear in a specifically human body, in specifically human flesh. Because, Athanasius once again points out, only human beings have gone astray. By implication, no other creature needs the incarnation of the Word in its own form. So the Word takes a part of the whole, a human body, as the instrument of human salvation. But this salvation is not limited to the human. In the human body of Jesus, God has touched every part of creation.[35]

Anselm (1033–1109)

Anselm was archbishop of Canterbury and undoubtedly one of the great theologians of the Middle Ages. His *Why Did God Become Human?* is conducted in the style of a dialogue between Anselm and the unfortunately named Boso. The opening question, posed by Anselm, is the decisive one: "For what reason and by what necessity did God become [hu]man?"[36] Note that Anselm's question alludes directly to the second Christological affirmation of the Creed: *homo factus est.* The soteriological agenda so apparent in Athanasius's work does not have to be argued here. It is simply assumed and left aside. The focus falls on the question of human nature and of God's assumption of it, as we see a little later when Boso reiterates Anselm's question in his own words: "By what necessity and for what reason God, although He is almighty, took on the lowliness and weakness of human nature, to restore it."[37] There is no emphasis, in the wording of this question, on human *flesh* or a human *body.* Human nature could exist, apparently, without reference either to flesh or a body. Again, much later, Anselm repeats that "Our inquiry is concerned only with the incarnation of God, and with what we believe regarding that humanity assumed by God."[38] Although he mentions the incarnation here, it is, as elsewhere in *Cur Deus Homo?*, the incarnation not of the second Person but of an undifferentiated God. Once again the emphasis falls on the assumption by this God of human nature rather than human flesh, far less flesh in general. To be fair, it has to be said that

35. Ch. 45 (PG 25, 176–177). Athanasius cites Colossians 2:15 here.
36. Bk. 1, ch. 1 (PL 158, 361).
37. Bk. 1, ch. 1 (PL 158, 362).
38. Bk. 1, ch. 10 (PL 158, 376).

Anselm had already written on the incarnation of the Word.[39] The earlier work, though more explicitly Trinitarian, is no less anthropocentric.

In Book 1, chapters 16–18, Anselm concerns himself with those particularly disembodied beings, the angels. The question that underlies this discussion is whether human beings, to be saved, must become "equal to the good angels."[40] It is the angelic rather than the embodied state that the Christian aspires to. But the passage that betrays Anselm's attitude to the natural world most tellingly comes in chapter 21. The exchange deserves quoting in full:

> ANSELM: God is in need of nothing, and if all things should perish, He could replace them, just as He created them.
>
> BOSO: I must acknowledge that I should do nothing against the will of God, to preserve the whole creation.
>
> ANSELM: What if there were more worlds than this one, full of creatures?
>
> BOSO: If the number of them were multiplied indefinitely . . . my answer would be the same.
>
> ANSELM: You could not be more right.[41]

We need read no further to see in Anselm a literary creator not only of the extraterrestrials of science fiction but also of a certain "apocalyptic fantasy" involving a cavalier disregard of the present, created universe.[42] Anselm's God has become human to rescue human beings from the sordid earthy reality in which they are embedded and to bring them, preferably without the rest of creation, to their true angelic home.

Comparison of the Two Samples

Athanasius and Anselm exemplify a profound difference in Christological emphasis. Athanasius is concerned with the Word, the second Person of the Trinity, who becomes flesh, becomes that which we human beings share with God's other animate creatures, the biblical *kol basar*. Certainly Athanasius moves on to relate this becoming flesh to the specifics of human flesh, but the point is that he holds together the two fundamental credal affirmations about the incarnation: "and became flesh . . . and was made human." In discerning this theme, I do not want to

---

39. *De fide Trinitatis et de incarnatione Verbi*, in PL 158, 259–284. Anselm refers to it in *Cur Deus Homo?*, bk. 2, ch. 9 (PL 158, 407–408), in what is the first explicitly Trinitarian reference in this work.

40. Bk. 1, ch. 19 (PL 158, 390).

41. PL 158, 394.

42. See Jürgen Moltmann, "God and the Nuclear Catastrophe," *Pacifica* (1988) 157–170.

credit Athanasius with our contemporary ecological consciousness. In any case we simply cannot know the intention of an ancient author, other than what we are told in the text itself. I do not even suggest that my reading of Athanasius is the most obvious or natural of possible interpretations.[43]

What I do want to suggest, however, is that we can find resources in this passage for developing an eco-Christology. I do not believe that we can find these resources in Anselm, and an ecotheological perspective may actually be subverted by Anselm's approach. He is concerned not with the Word, but in an effectively non-Trinitarian way, with an undifferentiated God, who becomes not flesh but human. Here the discussion has moved exclusively to the second part of the credal affirmation about Christ, namely, the words "and made human." This is clearly at the expense of any consideration of human fleshliness or embodiedness, Anselm even suggesting that the proper state of human beings may be, like that of the angels, disembodied.

Anselm's approach to Christology has in this regard been vastly more influential than that of Athanasius. Its origins may lie much earlier, conceivably in the Christological debates at Chalcedon with their focus on Christ's nature or natures, in abstraction from his human embodiedness.[44] It becomes a component, in due course, in the Enlightenment's disenchantment of nature, which brings with it an even greater disjunction between humanity and nature. Both these questions are beyond the scope of this chapter.

The problem for us is the effect that this approach has on our thinking about Christology and, more specifically, incarnation. I suggest that we instinctively hear the term "incarnation" as referring, not to enfleshment, but to becoming human. Evidence of this is the anthropocentric rather than incarnational or embodiment emphasis of the vast majority of even recent works on Christology. The absence of any significant ecological concern in these works, and the complementary tendency of ecotheology to avoid Christology, is hardly surprising. The story, however, is not quite as bleak as it may seem, and I want to point very briefly to some of the recent and notable departures from this broad majority tradition.

43. For example, when Athanasius speaks of Christ's death "clearing the air of the malignity of the devil and all demons," he may well have been thinking in dualist, up-down, and even instrumentalist categories, that is, the air is cleansed to ensure the progress of souls through the air to heaven. But the internal textual evidence offers no more support for this sort of reading than it does for an ecological reading, and in any case we have no way of knowing exactly what Athanasius had in mind apart from what he tells us himself.

44. The important contribution to ecotheology by Paulos Gregorios (*The Human Presence: An Orthodox View of Nature* [Geneva: World Council of Churches, 1978]) may at least in part have been aided by the author's Orthodox but non-Chalcedonian background. See pp. 8 and 84ff.

But first I must say more about the two suggested guiding or regulative principles for future explorations into Christology, for any Christology that is to be defensible on ecotheological grounds. These principles are suggested by the wording of the Nicene Creed in relation to the incarnation and also by the way Athanasius speaks about the incarnation. First, a major ingredient of any successful eco-Christology is that *it will hold together the affirmation that Christ has come as a human being with the prior affirmation that the second Person of the Trinity has become flesh.* In other words, it can be taken as a methodological principle for Christology that we will speak in an explicitly Trinitarian fashion of the second Person and will emphasize that this second Person has become flesh before it identifies this flesh more specifically as human flesh. To speak of this second Person, as Athanasius does, as Word or Wisdom, rather than using the more common designation Son, is also to open up a broader, less anthropocentric way of conceptualizing the identity of the second Trinitarian hypostasis, to say nothing of the specific, gendered implications of Son language.

A secondary, less Christological, but more anthropological principle can be derived from this discussion. When the two credal affirmations represent competing priorities, as they sometimes do between the concerns of humanity and the concerns of all flesh, then *specific human concerns will be considered only in the wider context of the concerns of all flesh.* The concerns of all flesh will take logical priority to the concerns of the human, as they do in the Creed, even though the concerns of humanity may be allowed, as they do in the Creed, to have the final say.

Attempts at the Retrieval of Eco-Christology

A number of scholars have worked hard to develop an ecologically sensitive Christology. A wisdom interpretation of Jesus, developed by Elizabeth Johnson, has been taken in an explicitly ecological direction by Denis Edwards.[45] Another perspective, starting from the Christological hymn of Colossians, focuses on the cosmic Christ and attempts to go beyond the exclusive link between Christ and human redemption. Developed by Joseph Sittler, it has been popularized by Matthew Fox and used consciously by Jürgen Moltmann as a link between Christology and ecotheology.[46] An ecologically focused embodiment Christology has been

45. Elizabeth Johnson, *She Who Is: The Mystery of God in Feminist Theological Discourse* (New York: Crossroad, 1992); Denis Edwards, *Jesus the Wisdom of God: An Ecological Theology* (New York: Orbis, 1995).

46. See S. Bouma-Prediger, *The Greening of Theology: The Ecological Models of Rosemary Radford Ruether, Joseph Sittler, and Jürgen Moltmann* (Atlanta: Scholars Press,

developed by Sallie McFague.[47] Growing out of a tradition of post-Reformation covenant theology is another contribution offered by Sue Patterson.[48] The *theologia crucis,* which was a major plank of Luther's reform movement, has been restated by Douglas John Hall and developed in the direction of ecotheology by Larry Rasmussen and more recently by Norman Habel.[49] Finally, there is the ecotheology of Christ's glory developed in Eastern Orthodox thought.[50]

All these recent attempts at eco-Christology conform to the proposed rules to greater or lesser extent. Denis Edwards's approach to Christology is clearly Trinitarian, the second Person being distinguished from God by the term "wisdom." This designation allows the author to draw on a biblical tradition that gives prominence to creation in all its diversity. Edwards also develops ethical criteria that make possible decisions about priorities, so that while all flesh is of intrinsic value, human needs take priority in certain circumstances. Jürgen Moltmann has developed a panentheism that gives prominence to what Karl Barth called Christ's third mode of being as the cosmic Christ, the *pantokrator,* the holder of all things.[51] Behind this Christology lie Moltmann's earlier concerns for both Trinitarian uniqueness and embodiment as "the end of all God's works."[52] But Moltmann insists that it is the human content of the story of the incarnation that gives God's enfleshment or embodiment its particularity.

Of all the writers mentioned, Sallie McFague takes the question of embodiment furthest with her notion of the world as God's body, though

---

1995); Matthew Fox, *The Coming of the Cosmic Christ: The Healing of Mother Earth and the Birth of a Global Renaissance* (Melbourne: Collins Dove, 1988); Jürgen Moltmann, *The Way of Jesus Christ.*

47. Sallie McFague, *The Body of God: An Ecological Theology* (London: SCM, 1993).

48. Sue Patterson, "Creation and Postmodernity," in *The Task of Theology Today: Doctrines and Dogma,* ed. V. Pfitzner and H. Regan (Adelaide: Australian Theological Forum, 1998) 84–111.

49. Larry Rasmussen, "Returning to Our Senses: The Theology of the Cross as a Theology for Ecojustice," in *After Nature's Revolt,* ed. D. Hessel (Minneapolis: Fortress, 1992) 40–56; Larry Rasmussen, *Earth Community, Earth Ethics* (Maryknoll: Orbis, 1996), especially 282–294; Norman Habel, "Key Ecojustice Principles: A *Theologia Crucis* Perspective," *Ecotheology* 5–6 (1998–1999) 114–125.

50. Paulos Gregorios, *The Human Presence,* 73–75; Geevarghese Mar Osthathios, *Theology of a Classless Society* (Guildford and London: Lutterworth, 1979) 78; Dimitrios, Patriarch of Constantinople, in *Orthodoxy and the Ecological Crisis* (Gland: Ecumenical Patriarchate/WWF, 1990) 1; Zizioulas, "Preserving God's Creation," 2.

51. Moltmann, *The Way of Jesus Christ,* 279. This notion of a "third way" of Christ's being may originate with Athanasius (see n. 27 above).

52. Jürgen Moltmann, *God in Creation: An Ecological Doctrine of Creation* (London: SCM, 1985) 244.

her book is less explicitly Trinitarian than some of the others.[53] Sue Patterson's Christology is explicitly Trinitarian and draws the notion of perichoresis into the material world, so that this material world is in turn drawn into the Trinitarian life of God. Our salvation, if it is to mean anything, must involve the salvation of all God's creatures. Rasmussen's contribution is to explore the ecological significance for Christ's suffering and death, a death that unites him not only with all living creatures but also with the earth itself. The theology of Christ's glory refuses to allow the resurrection to become an escape from this identification with the earth. Rather, it is the beginning of the glorification of all creation.

In addition to what they have in common, each of these approaches makes its own contribution to a developing eco-Christology. The wisdom theme draws on a profound and generously inclusive tradition within the biblical writings. The cosmic Christ theme makes clear the Christological content of the wisdom motif, pointing to the relationship of Christ to creation prior to his incarnation in time and place. The embodiment theme earths the cosmic Christ, dignifying not just the body of one man but every creaturely body individually, and collectively all flesh. The covenant theme returns our focus to the salvific reaching out of God toward us together with all creation. The ecotheology of the cross identifies Christ's incarnation with the pain and mortality that afflict all bodies, and systematically explores the links suggested in Genesis 2 between *adam, admah,* and *dammim.* The ecotheology of Christ's glory points to the inclusivity of God's eschatological reign, to which all things are lovingly invited.

Conclusion

Athanasius and Anselm represent two Christological approaches, two interpretations of the Creed, that can be seen throughout the history of Christian theology. Through a survey of their Christological reflections, we have discerned a growing disjunction of Christology from ecological concerns. This disjunction corresponds to the separation, in the central Christological affirmation of the Creed, of the two phrases "became flesh" and "was made human." To retrieve the nexus between these two phrases, I have suggested two principles for an earth-friendly Christology.

I have argued that the two proposed principles are to a greater or lesser extent implicit in all the above-mentioned recent attempts toward developing an ecologically sensitive Christology. These point to an interconnection between humanity and all flesh, and understand Christ to stand

---

53. This idea McFague adopts from Grace Janzen, *God's Body, God's World* (Philadelphia: Westminster, 1984).

in a further, Trinitarian interconnection that gives him his ultimate identity. The human story of Jesus Christ is not lost. In fact, it is essential if, as Moltmann has argued, the image of the cosmic Christ is to retain its concreteness and particularity. Salvation encompasses all creation but is mediated through the specifically human Jesus of Nazareth. As to why salvation needs a human mediator, Athanasius has offered a plausible reason: it is humans who have caused the problem.

As a step toward developing a truly earth-revealing, earth-healing Christology, my suggestion is that these principles be accepted as explicit guidelines to inform future Christological work. Christology can only start to become more ecologically sensitive when the human Christ is first enfleshed. The phrase "was made human" needs to be understood as an amplification of the earlier statement, "was incarnate," an amplification that refuses to supersede or override it but rather builds upon it and interprets it. The human is then understood as one particular way among others of being enfleshed. These guidelines may involve a reappraisal of some of the unquestioned assumptions of theological anthropology and call for the recovery of a much neglected strand in Christological thinking. Incarnation will be understood more literally as affirming the value of flesh in its broadest and most inclusive sense.[54] As neither biblical nor ecological consciousness can conceive of living flesh, whether human or not, in isolation from the waters, the dry land, and from plants yielding seed of every kind, this Christological affirmation of the value of flesh brings with it a logically prior affirmation of the interconnectedness of all things, animate and inanimate alike; or rather, it ratifies this affirmation already announced in the biblical creation stories. With the incarnation the relationship of creaturehood, in which all these things rejoice before God, has been deepened into a relationship of kinship with the eternal Word.

54. The term "incarnation" refers etymologically, after all, not to the human but to flesh *(caro/carnis)*.

Peace Bird

Land is good because there are a lot of trees
and birds. It gives bush tucker and good medicine.
The bush means good food, healing. Being in the
bush is more comfortable than the city, because
it's where I belong. Emu is a bird from my totem,
big mobs of emu. We do an emu dance, both men
and women dance it. We always ask permission
from old people to take others to the bush. Without
the land we have nothing.

"Peace Bird" is a black pigeon. Sometimes I put leaves
in its mouth, like the bird sent out (from the Ark).

Robert Burrinybuma

Chapter 5

God's Shattering Otherness

The Trinity and Earth's Healing

Patricia A. Fox

Australian social commentator and cartoonist Michael Leunig, reflecting on Australia's national identity, has this to say:

> We have damaged many ecologies and understood only when it is too late. Now we fancy that we are environmentally aware, but do we yet understand the ecology of human nature, the fragile ecology of the human soul? The spiritual eco-system upon which everything depends? . . . We flaunt our irreverence and religious impotence and call it enlightenment as we go in hot secular pursuit of material wealth and the fantasy that we are young and free and clever and changing and new and very reasonable. . . . In the proposed preamble to the Constitution, we seem prepared to honour the ancient aspects of indigenous culture, but nowhere in the document do we honour our own ancient culture. It seems that the indigenous people must now serve as tokens for the spiritual and the ancient on behalf of a society which cannot cope with the mystery of its own soul.[1]

It has been said that no nation on earth is as secular or irreverent as Australia, but I suspect that Leunig's observations hold some currency beyond these shores. The ecological crisis that humanity faces has been widely linked to religious belief and culture. In fact, Christian theology and practice have been named as contributors to a condition described as ecocide.[2]

---

1. Michael Leunig, "Denying Our Pain and Calling it Progress," *Sydney Morning Herald* (October 29, 1999) 15. Speaking from the perspective of a person of European descent, Leunig offered his reflections were made during the Australian debate leading to a referendum on changing the Constitution in favor of a republic.

2. See Elizabeth A. Johnson, *Women, Earth and Creator Spirit* (New York: Paulist Press, 1993) 10.

The discipline of ecotheology is one strand of Christianity's response to this global dilemma. It is an attempt to do theology from the perspective of the earth and is based on the premise that all of creation reveals the divine. It is a rereading of the Christian tradition that seeks to contribute to the earth's healing. In this chapter I want to go to a symbol that is at the heart of Christian culture and tradition and argue that the ancient symbol of God as Trinity discloses truths about the essential interconnectedness of both the fragile ecology of the human soul and of the planet we inhabit. These truths, I believe, can contribute to our whole ecology in ways that are liberating and life-sustaining.

In the early Church, nature, along with God and humanity, was recognized to be one of the three pillars of theology.[3] Later, in many places, physical matter ceased to be celebrated as God's gift. Nature came to be rejected as insignificant, or worse, in some instances it came to be regarded as a source of evil.[4] It became an *other* to be feared and conquered. It was considered dangerous for those genuinely seeking God. During the post-Enlightenment period in the West, non-human nature was treated as a commodity to be harnessed for humanity's use. This attitude prevailed through the industrial revolution and into the modern era and has been identified as a powerful contributing factor to the ecological crisis we face now.

In these times there has been a major shift in consciousness. As theologian David Tracy observes, the modern "turn to the subject" has been replaced in much contemporary thought by the postmodern "turn to the other":

> The real face of our period, as Emmanuel Levinas saw with such clarity, is the face of the other: the face that commands, "Do not kill me." The face insists: do not reduce me or anyone else to your narrative. . . . God's shattering otherness, the neighbor's irreducible otherness, the othering reality of "revelation" (not the consoling modern notion of "religion"): all these expressions of genuine otherness demand the serious attention of all thoughtful persons.[5]

Tracy suggests that, in our times, it is the presence of massive global suffering that has brought to center stage "the reality of the others and the different."[6] God above all is other and different. And as Tracy describes,

---

3. See Elizabeth A. Johnson, "Turn to the Heavens and the Earth: Retrieval of the Cosmos in Theology," *Catholic Theological Society of America Proceedings* 51 (1996) 1–14.

4. I am describing in broad strokes a story of nature within the Western European experience that has shaped much of Christian theology. With this dominant movement there coexisted alternative traditions. See, for example, Rosemary Radford Ruether, *Gaia and God: An Ecofeminist Theology of Earth Healing* (New York: HarperCollins, 1992); Matthew Fox, *Original Blessing* (Santa Fe, N.M.: Bear & Co., 1983).

5. David Tracy, "The Hidden God: The Divine Other of Liberation," *Cross Currents* 46 (1996) 5–6.

6. Ibid., 7.

"God's shattering otherness" also shouts for recognition. This is not a divine otherness of distance and disinterest but an otherness or difference whose essential characteristic is to be in relation.

In this chapter I want to give serious attention to God as *other* and to non-human nature as *other*. I want to listen to their cries: "Do not kill me!" I want to do it in a way that does not reduce either to humanity's narrative but rather acknowledges that God, human nature, and non-human nature must be considered in mutual relation to each other. In order to do this, I will draw from the work of two contemporary theologians, John Zizioulas, a metropolitan of the Greek Orthodox Church, and Elizabeth Johnson, a Roman Catholic feminist theologian. It is no accident that both these theologians have made significant contributions to the retrieval of the central Christian symbol of the Trinity *and* have also seriously addressed the ecological crisis. Both argue, albeit for different reasons, that if the issue of global destruction of the earth is to be resolved it must be considered in relation to the question of God.

Contemporary scholarship suggests that very soon after the doctrine of the Trinity was formulated, theological discourse began to separate the triune God of salvation history from the trinity of Persons within God. A living Trinitarian spirituality that had evolved from the eucharistic experiences of the early Christian communities became disconnected from its biblical base. Instead, Trinitarian reflection gradually became centered within the esoteric realms of speculative theology.[7] By the medieval period in both Latin and Byzantine theology, the doctrine of the Trinity was often understood to refer to the inner life of God (the immanent Trinity), with little reference to God's deeds in history (the economic Trinity). This has meant that for a millennium and a half, the doctrine of the Trinity has tended to be restricted to consideration of God in Godself. The theology of the textbooks created the impression that creation has no essential relationship with the Trinity and vice versa. In fact, by this century Karl Rahner could make his oft-quoted remark that one could dispense with the doctrine of the Trinity and the major part of religious literature would remain virtually unchanged.[8]

Thus, despite the fact that Christianity has preserved a strong tradition of belief in the triune God through its liturgy and dogma, a powerful monotheistic image of God as an omnipotent, omniscient male monarch

---

7. See Catherine Mowry LaCugna, *God for Us: The Trinity and Christian Life* (New York: HarperSanFrancisco, 1991) for one description of this process and its consequences. For a critical appraisal of LaCugna's thesis, see Duncan Reid, "The Defeat of Trinitarian Theology: An Alternative View," *Pacifica* 9 (1996) 289–300.

8. Karl Rahner, *The Trinity* (New York: Herder & Herder, 1970) 11.

has shaped both popular devotion and the discourse of Christian Churches into the modern era. It is only within the last decades of the twentieth century that a significant alternative direction has reemerged within theology. There has been a movement toward a retrieval of the Holy Mystery of the triune God, who was foreshadowed in the Scriptures, claimed in Christian doctrine, and constantly confessed in public worship throughout the ages.

Contemporary theologians have returned to the origin of the symbol's formulation. The revelation of the God proclaimed by Jesus of Nazareth was heralded by early bearers of the Christian gospel to be "a teaching that is new" (Mark 1:27). Former categories for understanding divinity were exploded by the God revealed to those initiated into the life of the early Christian communities. Here it was claimed that the great, personal, and saving God of Israel had become irrevocably joined with humanity in the person of Jesus and continues to be present in the risen Christ through the power of the Spirit. This had radical implications for the way Christians viewed matter. It was believed that "in Christ" all creation is destined for eternity by being drawn into the communion of God's own life.[9]

Today we inherit new understandings of relation and person along with a thrilling and still unfolding, new story of the origins and nature of the cosmos. The recent renaissance in Trinitarian theology demonstrates that because of these new frames of reference, the Trinitarian doctrine of God can function in fresh ways to reveal truths about the God of Jesus Christ. And, more significantly, given the massive ecological crisis confronting humanity, Christian theology is seeking to discover what this central Christian symbol can mean for human and non-human creation today. It is seeking to retrieve and communicate the *otherness* of the triune God that still shatters the idolatrous boundaries of humanity's limited constructs about who God is.

The very different theologies of Elizabeth Johnson and John Zizioulas both link the retrieval of God as Trinity with the restoration of non-human nature to its proper place in the sphere of the sacred.[10] Zizioulas's theology is essentially ecclesial, while Johnson's is from the perspective of life in the world. Both engage in a historical analysis of Christianity's involvement with the crisis, but Zizioulas's theological contribution to its solution is drawn primarily from the Eastern tradition and is eschatological and liturgical. Johnson's contribution is from a Western

---

9. See Chapter 7 by Anthony Lowes in this volume for a much fuller elaboration of the Christian belief that "in the end, matter matters."

10. See Patricia A. Fox, *God as Communion: John Zizioulas, Elizabeth Johnson and the Retrieval of the Symbol of the Triune God* (Collegeville, Minnesota: The Liturgical Press, 2001).

feminist liberation theology that is based on women's experience and is directed to a just praxis. Both of their theologies are anchored in an understanding of the radical otherness of God and open up two powerful and complementary visions of the Trinity as divine *koinonia* (communion). Together they provide a glimpse of how reclaiming the triune symbol of God can have profound existential consequences. I will focus briefly on some key elements of their respective theologies and show how these contribute to an ethos and an ethic for a restored ecology.

The Trinitarian God as "Persons in Communion"

At the core of John Zizioulas's Trinitarian theology is his vision of God as "Persons in communion." His research into the patristic sources of the doctrine of the Trinity reveals that it was the struggle in the first centuries of the Church to develop language to speak about the Christian experience of the triune God of love that generated a breakthrough in the ontology of person.

Zizioulas claims that it was the experience of living and worshiping in Christian communities that revealed something very important about God. It revealed that the being of God came to be known through life in community, through personal relationships and personal love, and that "Being means life, and life means *communion.*"[11] This experience of living in the Church as communion *(koinonia)* provided an existential reference point for those leaders (bishops) who were working on formulations of the doctrine of the being of God. Zizioulas refers specifically to Athanasius of Alexandria and the Cappadocians, Basil the Great, Gregory of Nazianzus, and Gregory of Nyssa. The key question that they had to address was: "What does it mean to say that God is Father, Son and Spirit without ceasing to be *one* God?"[12] Following the Hebrew tradition, Christianity was able to engage with this question because it held that *being,* the existence of the world, was created out of nothing and was therefore a product of freedom. Christian patristic thought was not bound by the ontological necessity of Greek philosophy. Moreover, a biblical world picture held that the basis of this freedom was the being of a God who is *personal.*

---

11. John D. Zizioulas, *Being as Communion: Studies in Personhood and the Church* (New York: St. Vladimir's Seminary Press, 1993) 16–17.

12. See John D. Zizioulas, "The Contribution of Cappadocia to Christian Thought," in *Sinasos in Cappadocia,* ed. Frosso Pimenides and Stelios Roïdes (National Trust for Greece: Agra Publications, 1986) 25. Zizioulas refers here to "a whole series of Christian thinkers" who addressed this question, including Irenaeus, Clement, Origen, and Athanasius.

Drawing from patristic sources and particularly from the Cappadocians, Zizioulas argues that the being of the triune God *is* the communion between the three Persons. He describes being as communion and holds that there is no true being without communion. "Person" therefore can be conceived, not as an adjunct to being, but as being itself. To be a person means to be in relation. Zizioulas focuses on relation understood as an ontological category and argues that in God it is possible for the particular to be ontologically ultimate because of the unbreakable and living communion between Father, Son, and Spirit. He emphasizes that God's love, which came to be identified with ontological freedom, resides, not in God's nature, but in God's personal existence.[13]

From his probing of the origins of Trinitarian doctrine, Zizioulas concludes that personhood is human existence's "most dear and precious good."[14] He argues that the concept of relation has vital relevance for contemporary studies of personhood and for addressing the roots of the ecological crisis.[15] He traces the historical evolution of the concept of person within Western thought and claims that it was derived from a combination of two basic sources: *rational individuality* (Boethius) and *psychological experience and consciousness* (Augustine). Thus, "person" came to be understood as an *individual:* "a unit endowed with intellectual, psychological and moral qualities centered on the axis of consciousness . . . an *autonomous self* who intends, thinks, decides, acts and produces results."[16] In stark contrast to this understanding of person, Zizioulas wants to reclaim the patristic concept that developed from within Trinitarian theology. He claims that:

> being a person is basically different from being an individual or "personality" in that the person cannot be conceived in itself as a static identity, but only as it relates to. Thus personhood implies the "openness of being," and even more than that, the *ek-stasis* of being, i.e. a movement towards communion which leads to a transcendence of the boundaries of the "self" and thus to freedom.[17]

This is an extraordinary claim from theology for the meaning and potential of "person" as a relational category. Zizioulas notes that such a claim

13. See Zizioulas, *Being,* 41–47.

14. Ibid., 65.

15. See John Zizioulas, "Preserving God's Creation. Three Lectures on Theology and Ecology," *King's Theological Review* 12 (1989) 1–5, 41–45; 13 (1990) 1–5.

16. John Zizioulas, "Human Capacity and Human Incapacity: A Theological Exploration of Personhood," *Scottish Journal of Theology* (1975) 405–406.

17. Ibid., 407–408. Zizioulas notes that the term *ek-stasis* used here is known mainly through the philosophy of Martin Heidegger but that the term was in fact used before him in patristic mystical writings (e.g., Pseudo-Dionysius and Maximus the Confessor).

is in harmony with other contemporary understandings of person.[18] Chapter 11 in this volume, by James McEvoy, elaborates the important contributions to this area of discourse by the philosopher Charles Taylor.

Zizioulas develops the connections between relation and personhood further by insisting that person cannot be understood only as the "ecstasy" of substance, by which he means "a movement towards communion which leads to transcendence of the boundaries of the 'self' and thus to freedom."[19] He argues that it must also be understood as a *hypostasis* of substance, that is, as a particular identity. Thus, for him, *"ekstasis* and *hypostasis* represent two basic aspects of Personhood . . . the idea of Person affirms at once that being cannot be 'contained' or 'divided', and that the mode of its existence, *hypostasis,* is absolutely unique and unrepeatable."[20]

It is significant to note here that Zizioulas is asserting, as much as the Western thinkers that he critiques, a uniqueness or individuality for personhood under the name of *hypostasis*. The difference is that he holds that the particularity or hypostasis is constituted only through the relationality.[21] True communion only occurs when the uniqueness of each entity is fully respected. In turn, it is only in communion that each entity can come to its full, unique potential. This is a key point in his elaboration of the concept of person and in his approach to addressing the massive challenge of the ecological crisis.

Zizioulas further emphasizes that theology, unlike philosophy, teaches an ontology that ensures the absolute and unique identity beyond even death. For all created entities, death overcomes communion because relationships must end with death. He argues, however, that because life in God is personal—in communion—life is eternal. In the realm of the uncreated, within the triune God, life and communion coincide:

> The eternal survival of the person as a unique, unrepeatable and free "hypostasis," as loving and being loved, constitutes the quintessence of salvation

---

18. See Zizioulas, "Human Capacity," 408, where he mentions some representative examples of understandings of the person as a relational category—M. Buber's *I and Thou;* J. Macmurray's *Persons in Relation* and *The Self as Agent;* W. Pannenberg's article "Person," *Religion in Geschichte und Gegenwart V*, 3rd ed. (Tubingen: J. C. B. Mohr, 1958) 230–235.

19. Zizioulas, "Human Capacity," 408.

20. Ibid.

21. For a feminist elaboration and example of the phenomenon that Zizioulas describes here, see Paula M. Cooey, "Emptiness, Otherness and Identity," *Journal of Feminist Studies in Religion* 6 (1990) 20–21. Reflecting on her own experience, Cooey writes: "The connection I experience is not one of fusion, but one of communion in which the other and I remain distinct . . . other is present to me—present as distinctly other [and] positively grants me my own distinctiveness."

. . . this is called "divinisation" *(theosis)*, which means participation not in the nature or substance of God, but in His personal existence.[22]

Thus he comes to the significant conclusion that "salvation is identified with the realisation of personhood" and that coming to complete fullness can only happen in a final way in God *(theosis)*.[23] He is saying that to become fully a person, ecstatically and hypostatically, is to break through the isolating boundaries of individualism into a life of inclusive communion with persons and with all creation.

Zizioulas claims that this essentially theological understanding of person resulted from a breakthrough that occurred through the Christian struggle to speak about the God revealed by Jesus Christ. He further claims that this has immense practical and existential implications for contemporary life. And clearly it does. If it were a universally held truth that becoming fully a person in love and freedom means breaking through the narrow boundaries of individualism into a life that values difference in communion, the world would be a vastly different place. Such an understanding of person would radically affect how the peoples of this planet could live together and how they would together live in relation to all of creation.

It is perhaps within the context of the Christian eucharistic liturgy that a Westerner can best begin to glimpse the implications for creation of Zizioulas's imaging of God as "Persons in communion." He draws attention to the early Christian awareness that it is the Holy Spirit, the Life-giver, the One who is the source of the many charisms in the community, who gathers the diverse members into assembly. When the different persons of the community come together, they bring with them the gifts of creation. They bring bread and wine to represent themselves, their lives. By offering to God these gifts of creation, those assembled symbolically acknowledge their unity and identity with creation. At the one table of the eucharistic liturgy the *many*—the gathered community joined with the rich diversity of the whole of creation—are constituted by the Spirit into the *one* Person of Christ. It is as the body of Christ that the *many-become-one* are offered to God, the Source of all being, and are drawn into the communion of the triune God. Entering into the liturgical action thus provides a way for the believer both to relate to each of the three Persons in God and to participate in the dynamic relationality and love of the three Persons in communion. Zizioulas is referring to this experience when he states that "the existence of God is revealed to us in the Liturgy as an event of communion."[24]

22. See John Zizioulas, "The Teaching of the 2nd Ecumenical Council on the Holy Spirit in Historical and Ecumenical Perspective," in *Credo in Spiritum Sanctum,* ed. J. S. Martins (Rome: Libreria Editrice Vaticana, 1983) 1:53; Zizioulas, *Being,* 49–50.

23. Zizioulas, *Being,* 50.

24. Zizioulas, "2nd Ecumenical Council," 39.

Thus Zizioulas's return to the origins of the Trinitarian formulation of God as three Persons in mutual and equal relation recovers a key understanding that *the being of God is communion*. This is a foundational insight. It addresses the problem of Western theology which speaks of God's being as a substance and which has served over the centuries to communicate that God is remote from the concerns of creation. The symbol of God as "Persons in communion" communicates the strong biblical teaching that God is personal and also that to be a person is to be in dynamic mutual relation with all other persons and created entities. God, imaged as different Persons in communion, reveals uniqueness flourishing through mutual relationships. This understanding of God values difference and allows for the intrinsic value of every created entity. It can accommodate the new story of the universe. At the same time, this Trinitarian picture of one-God-who-is-three also creates an immediate cognitive dissonance. It cannot be easily imagined. It does not "fit." It shatters human categories. God is other than we think! Thus, while communicating essential elements of the biblical revelation of God, it also functions to communicate God as the eschatological Other, the one who ultimately transcends all description.

The Trinitarian God as "She Who Is"

Elizabeth Johnson's Trinitarian theology is based on the premise that a retrieval of the doctrine of God can serve a discourse about divine mystery that will further the emancipation of women.[25] The task she sets for herself in her major work on the Trinity is to probe Christianity's inherited tradition in order to "accomplish a critical retrieval [of a core religious symbol of God] in the light of women's coequal humanity."[26] Then, in a further critical move, she links the neglect of the Spirit with the exploitation of the earth and the marginalization of women and argues that because these three relationships are so intrinsically interconnected, none of them can be addressed in any useful way in isolation from each other. She holds that we have lost a sense of the sacredness of the earth and that this is "linked to the exclusion of women from the sphere of the sacred, which is tied to [a] focus on a monarchical, patriarchal idea of God and a consequent forgetting of the Creator Spirit, the Life-giver, who is intimately related to the earth."[27] Her analysis is that the Holy Spirit can only be effectively reclaimed within Trinitarian theology and be fully reinstated

---

25. See Elizabeth A. Johnson, *She Who Is: The Mystery of God in Feminist Theological Discourse* (New York: Crossroad, 1992) 9.

26. Ibid., 10.

27. Johnson, *Women, Earth and Creator Spirit*, 21.

within the Christian life if the major issues of ecology and sexism are simultaneously addressed.

Many theologians, and particularly feminist theologians, have identified the power of language for naming God as a critical issue. Johnson's question concerning the right way to speak about God can be situated within the rising concern of all people who have begun to recognize the profound implications of speech about God both for the future of life on this planet and for the human person's capacity to know and relate to God.[28] Respect for God's absolute otherness ensures that any name that is used for God has, by necessity, only a limited currency. If idolatry is to be avoided and if human persons are to be open to the full revelation of who God can be for humanity and for the whole of creation, many names are essential. Johnson reasons, therefore, that any serious retrieval of the symbol of the triune God can only properly occur when the male imagery that has been used almost exclusively of the triune God for two thousand years is acknowledged as problematic and is addressed.

However, because naming the three Persons must be congruent with the biblical witness and communicate their relation of origin with each other, the task of using many names for the Trinity poses particular difficulties. Johnson's feminist theology of the Trinity tackles this challenge very directly. She renames God as SHE WHO IS, and she renames and reorders the divine Persons as Spirit-Sophia, Jesus-Sophia, and Mother-Sophia. This renaming and reordering provide the foundation for an innovative and constructive proposal based on a feminist re-reading of the tradition.

Since it is possible to explore only a couple of the elements of this proposal within the scope of this chapter, I will focus on the name SHE WHO IS and, within this naming, Johnson's redefinition of divine omnipotence. Both of these have a direct bearing on this theologian's response to the critical global issue that she describes as ecocide.[29]

Johnson entitles one of the chapters of her book on the Trinity "Triune God: Mystery of Relation." She acknowledges that Trinitarian theology, founded as it is in the biblical story of salvation, has consistently taught that what constitutes the three divine Persons is their relationality; that relationship both constitutes each Trinitarian Person as unique and distinguishes one from another. It is on the basis of this teaching that she describes the Trinity as "a mystery of real, mutual relations." In fact, she focuses on the belief that "at the heart of holy mystery is not monarchy but community; not an absolute ruler, but a threefold *koinonia*."[30] She notes that this points to a

---

28. For example, see Sallie McFague, *Models of God: Theology for an Ecological Nuclear Age* (London: SCM Press, 1987).

29. See Johnson, *Women, Earth and Creator Spirit*, 10.

30. Johnson, *She Who Is*, 216.

sense of ultimate reality that is very congruent with the feminist values of mutuality, relation, equality, and community in diversity.[31]

Johnson's attempt to articulate the "radical livingness of God's holy mystery" leads her to an exploration of God's being. In expressions remarkably similar to those used by Zizioulas, she acknowledges that "divine unity exists as an intrinsic *koinonia* of love" and that "being in communion constitutes God's very essence."[32] Building on this premise, she seeks to use the ontological language of being to new benefit:

> I suggest that the ontological language of being has the advantage of providing an all-inclusive category for reality at large, leaving nothing out and thereby entailing that the cosmos does not slip from view by too heavy a concentration on the human dilemma. . . . It is thus a code word for God as source of the whole universe, past and present, and yet to come, and as power that continuously resists evil.[33]

It is this notion of being that evokes "a most dynamic and living although elusive reality, the act of being-there of things."[34] It provides an all-inclusive category for reality and, as Johnson notes, functions to bring the cosmos into view alongside humanity. Tuning directly into the power of Thomas Aquinas's thought, she suggests that the language of being so understood can communicate that "all things are on fire with existence by participation in God's holy being which is unquenchable."[35] When applying the language of being to God, Johnson reminds the reader that "the being of God that we are speaking of is essentially love. God's being is identical with an act of communion, not with monolithic substance, and so is inherently relational."[36]

It is through this lens of being that Johnson focuses on the powerful story of encounter between Moses and God described in the Book of Exodus. The story tells of a compassionate God who has heard the cry of the enslaved Israelites, who "knows" their sufferings and seeks to deliver them into freedom (Exod 3:7-8). And the self-identifying name that God gives on this occasion to the bewildered Moses is YHWH, I AM WHO I AM. Johnson notes that biblical commentators are unanimous in their criticism

---

31. See ibid., 211.

32. Ibid., 227–228.

33. Ibid., 237.

34. Ibid., 238.

35. Ibid.

36. Ibid. For another contemporary retrieval of Aquinas's concept of being, see Anthony Kelly's transposition of divine "Being" to "Being-in-Love," elaborated in Anthony Kelly, *The Trinity of Love* (Wilmington, Del.: Michael Glazier, 1989) and in Anthony Kelly, *An Expanding Theology: Faith in a World of Connections* (Newtown, N.S.W.: E. J. Dwyer, 1993).

of the anachronistic tendency to read a philosophical meaning back into this text. However, she also notes that from the time of the Septuagint translation onward, the idea that the name YHWH discloses the ontological nature of God gained credence in Jewish circles and was widely used in early Christian theology.[37]

Thus Aquinas was drawing from a revered ancient tradition when he sought to support his claim that God's very nature is *to be,* by consciously making a connection between philosophy and Scripture and appealing to a metaphysical interpretation of this text from Exodus. He sought to communicate a God of pure transcendent being in whom the whole created universe participates and argued that the name QUI EST is the most appropriate name for God.[38] Johnson observes that this Latin construction can be literally translated as "who is" or "the one who is." She therefore argues that since God is not intrinsically male, this highly influential text, which carries the meaning of a divine relational being who energizes the world, could be translated with a feminist gloss as "SHE WHO IS." Moreover, because this name is suggestive of the God revealed in the narrative of the burning bush, it also functions to bring to mind a personal, compassionate, faithful God who sustains a loving relation with all of creation.

> Symbolized by a fire that does not destroy, this one will be known by the words and deeds of liberation and covenant that follow. SHE WHO IS, the one whose very nature is sheer aliveness, is the profoundly relational source of being of the whole universe.[39]

Thus the biblical story of liberation from Exodus is key to Johnson's proposal to name the triune God SHE WHO IS. Quoting J. B. Metz, she recalls that "the pure idea of God is in reality, an abbreviation, a shorthand for stories without which there is no Christian truth in this idea of God."[40] This name then, as an abbreviation of the saving story of Exodus, can powerfully evoke the mystery of God in solidarity with the oppressed. It can convey the mystery of a God who is dynamically present to the needy and who is active to free all that is bound. Moreover, by connecting with the aspect of the story where a reluctant Moses is enlisted into a collaborative work of liberation, naming God as SHE WHO IS also conveys a call to humanity for mutuality with her in the task of saving the world. Thus:

---

37. See Johnson, *She Who Is,* 242. See also p. 302, n. 38, where Johnson refers to Walter Kasper, *God of Jesus Christ,* trans. Matthew J. O'Connell (New York: Crossroad, 1991) 147–152, for a history of how the name became a metaphysical definition.

38. See Thomas Aquinas, *Summa Theologiae,* I, q. 13, a. 11.

39. Johnson, *She Who Is,* 243.

40. Ibid., 244. See Johannes Baptist Metz, "Theology Today: New Crises and New Visions," *Catholic Theological Society of America Proceedings* 49 (1985) 7.

> . . . alive in the *koinonia* of *SHE WHO IS,* women and men are called to be friends of God and prophets. . . . This way of speaking crafts a partnership amid the ambiguity of history: *SHE WHO IS,* Holy Wisdom herself, lives as the transcendent matrix who underlies and supports all existence and potential for new being, all resistance to oppression and the powers that destroy, while women and men, [like Moses] through all the ambivalence of their own fidelity, share in her power of love to create, struggle, and hope on behalf of the new creation in the face of suffering and evil.[41]

Naming the one God SHE WHO IS thus communicates God's dynamic being as *koinonia,* as a mystery of relation at the heart of the universe: it conveys the image of a God who dwells within, giving life to all that is. Conversely, it also communicates the image of a powerful personal female presence who makes room for a dynamic, real, and reciprocal relation with creation: SHE WHO IS evokes the image of a God in whom all creation lives and dwells and has its being. Finally, it also discloses the faithful God of the covenant as Holy Wisdom, a compassionate God who identifies with suffering creation and who draws humanity to join her in the work of the liberation of all creation from suffering and evil. SHE WHO IS invites each person to action on behalf of justice. This vision of God, calling to mind as it does the story of Moses paralyzed in the face of suffering, presents a picture of a God who persists in the task of engaging humanity in her work of healing. This includes, in Johnson's rereading of these texts, the Trinitarian God drawing humanity into the work of healing planet Earth.

Central to any work of healing is the capacity to face suffering.[42] A significant component of Johnson's struggle "to speak rightly of God" springs from her acknowledgment that one of the major challenges to theology in this century is the massive and acute presence of suffering in the world.[43] She argues that "anything we say of God has to ring true in word and deed in the face of this pain."[44] She acknowledges that for many, the all-powerful, monarchical God who for centuries has dominated Christian theology and devotion is "dead." The image of God as a spectator to suffering simply did not ring true with the God of compassionate love revealed in the Hebrew Scriptures or with the God revealed by the life, death, and resurrection of Jesus of Nazareth. Johnson observes that this image of a

41. Johnson, *She Who Is,* 244–245.
42. See Lorna Hallahan's "Embracing Unloveliness: Exploring Theology from the Dungheap," Chapter 6 in this volume.
43. Ibid., 3.
44. Elizabeth Johnson, "Review Symposium, Author's Response," *Horizons* 20 (1993) 344.

God who is incapable of suffering enshrines unilateral power as an ultimate goal and derides vulnerability as weakness and imperfection. She suggests that this has begun to be replaced by images of a God of relational power that are congruent with scriptural and doctrinal witness.

Relational power stands in direct contrast to the unilateral power communicated by the omnipotent God of classical theism. It is founded on the liberating strength that emanates from the free capacity to choose to become vulnerable in order to sustain a mutual loving relationship—"the liberating power of connectedness that is effective in compassionate love."[45] This is the power revealed by a Trinitarian God of mutual relations whose very being is communion. It is not a power that has to control. It is a power that emanates from persons who are free to love. It is free to be vulnerable to the limits of the other and to let new life come into being as it will. It is a power that is free enough to enter into suffering. Since love entails suffering, a retrieval of the symbol of the triune God who is Love must include images of a God who suffers.[46]

To speak of a suffering God subverts the patriarchal image of perfection and the consequent ideal of unilateral power. It communicates that self-containment and the absence of relationship do not represent an ideal but rather signify imperfection. Omnipotence can thus be redefined as the free, unlimited capacity to make room within the self for the other. This evocative image of making room within oneself for another is one of the reasons that Johnson proposes that the name Mother-Sophia provides an appropriate alternative to naming one of the Trinitarian Persons exclusively as Father. This redefinition of omnipotence also leads her to propose that the Trinitarian God be called Suffering God: Compassion Poured Out. All the above insights toward a redefinition of power are congruent with the scriptural revelation of God. They need to be taken into account if the energy of the triune symbol of God is to be fully released in this era.

Johnson finally warns that the rule of analogical speech must be applied here with full rigor. She recalls that the language of God as suffering is not, of course, a literal description of God. Nevertheless, she has demonstrated that one can speak of a suffering God by reentering the powerful biblical story of covenant and liberation and remembering the living God—SHE WHO IS—revealed as being dynamically present with her suffering people and, by extension, her suffering creation. She shows that the God who suffers is also revealed by the redemptive image of Jesus-Sophia,

---

45. Johnson, *She Who Is,* 270.

46. See also Jürgen Moltmann, *The Crucified God: The Cross of Christ as the Foundation and Criticism of Christian Theology,* trans. R. A. Wilson and John Bowden (New York: Harper & Row, 1973).

the crucified one, love poured out; by Spirit-Sophia who, according to Scripture and rabbinic writing, is compassion poured out, forever "speaking, crying, admonishing, sorrowing, weeping, rejoicing, groaning, comforting";[47] by Mother-Sophia, the revelation of the fecundity of relational power and love who freely makes room within herself for creation to come to be. Johnson thus shows that to contemplate the mystery of the one Trinitarian God as a living mystery of personal relations at the heart of the universe is to come to know Holy Wisdom, the triune God. This one-God-who-is-three suffers with us and prompts us to ethical action.

Ecotheology and Healing the Earth: The Triune Symbol as Source for a Transforming Ethos and Ethic

Drawing from Zizioulas and Johnson, I suggest that there is a link between humanity's tendency to reduce both divine and non-human life to our narrative. A recognition and reception of "God's shattering otherness" have an intrinsic connection with our capacity as Christians to recognize and receive non-human nature as the *other*. In this last section I will focus on the significance of the triune symbol of God as a source for both an ethos and ethic that can create and sustain the commitment necessary to act in an enduring way on behalf of justice for the earth. Zizioulas spells out how a living spirituality of the Trinity nurtures such an ethos and enables changed attitudes and action toward the world. Johnson stresses the significance of a unifying theological vision that can provide a basis for the collective will necessary for the ethical action required for such a change.

Zizioulas argues that the ecological crisis is a crisis of culture. He describes it as "a crisis that has to do with the loss of the *sacrality* of nature in our culture."[48] While he acknowledges the value of the myths and rituals of many religions because they contribute to the development of an ethos that promotes the sacrality and therefore the survival of nature, he insists that Christianity can offer much more.[49] It teaches that when "the Word became flesh," uncreated and created became irrevocably joined in Christ. That means that what was destined to extinction is now drawn into

---

47. Johnson, *She Who Is*, 266. Johnson refers to Pope John Paul II, who describes the Spirit personifying the suffering God (albeit with a distinct theology of suffering) and entering "into human and cosmic suffering with a new outpouring of love, which will redeem the world." *Lord and Giver of Life (Dominum et Vivificantem)* (Washington, D.C.: United States Catholic Conference, 1986) no. 39.

48. Zizioulas, "Creation 3," 5.

49. See A. R. Peacocke, *Creation and the World of Science: The Bampton Lectures, 1978* (Oxford: Clarendon Press, 1979) 270ff. for a discussion of the connection between theology and ecological values.

relationship with the divine for eternity. Because of the incarnation, humanity is called to live in such a way that its deeply loving relationship with creation enables non-human nature to be drawn into divine communion. Zizioulas believes that in order to sustain such a calling, such a way of life, humanity needs the kind of ethos that will enable each person to become a "liturgical being."[50]

Central to Zizioulas's theological engagement with the issue of the ecological crisis is his understanding of "offering up" as a significant aspect of priesthood. While in one sense he is referring to a specific action of the ordained minister within the eucharistic liturgy, he is essentially referring to a priesthood that is to be exercised by every person. He holds that the Person of Christ as "Priest of creation" provides a model for humanity's proper relationship with the world.[51] Every person is to participate in the priestly task of "offering," of living and relating to all creation in such a way that each particular being is in turn opened up to a transcending relatedness with the "other," this action "corresponding to that of *love* in the deepest sense."[52]

This description of a "turn to the other" surely relates to the postmodernist instinct that Tracy refers to above. Zizioulas links it here to the action of the anaphora within the Eucharist, to the lifting up of creation to God the Creator that is to be played out continually around the planet in the life of every person.[53] In so doing, every person is initiating a priestly action whereby creation, having been offered to God, can be transformed by the action of the Holy Spirit and be brought to a fullness beyond its natural capacities. This is the sense in which the human being has to become a liturgical being so that the ecological crisis can be overcome.

According to Zizioulas, this is also a description of how humanity and all creation are drawn into the hypostatic and ecstatic life of the Trinity. The gathering in community to enter into the great mystery of eucharist—of thanksgiving for creation and salvation in Christ and the Spirit—becomes a transforming moment. Within this moment, each person's incorporation into the very life of God is enacted before being spun out again to continue this saving, relational, and priestly action in the world. Thus there is a direct connection between Zizioulas's understanding of the process of our becoming

50. See Zizioulas, "Creation 1," 2.

51. Zizioulas, "Creation 3," 5.

52. Zizioulas, "Creation 1," 2.

53. In Chapter 7 of this volume, "Up Close and Personal: In the End, Matter Matters," Anthony Lowes refers to the same phenomenon when he speaks of human persons contributing to the ontological transformation of "their habitat, their portion of the universe . . . to the degree that they have managed to love them unconditionally and serve them selflessly."

persons and his understanding of humanity's role in overcoming the ecological crisis. A person comes to be "in relation." A failure to relate lovingly to created entities diminishes a human being's capacity for personhood.

Zizioulas's engagement of theology with ecology in order to address the ecological crisis leads inexorably, through an articulation of a theology of creation, to the heart of all that is—to a God revealed as a communion of three divine Persons into whom humanity and all creation are drawn.

> Thanks to the economy of the Holy Trinity which has been realized in the person and the work of Christ, "with the cooperation of the Holy Spirit," space and time are capable of receiving transfiguration; . . . the Kingdom of God is not something that will displace material creation, but will transfigure it, cleansing it from those elements which bring about corruption and death. The Eucharist gives us the assurance that matter is sacred and worthy of every honour.[54]

The triune God is the dynamic ground and underlying premise of his proposal. Zizioulas is thus steadfast in demonstrating an intrinsic and ultimate relatedness between the Trinity, humanity and all of creation, and in claiming that this relatedness has very practical as well as ontological significance.

Zizioulas seeks to promote this whole liturgical vision as a "remedy of immortality."[55] The contemporary crisis he is addressing issues from stark, destructive existential realities: water, air, and the very earth itself are being destroyed; species of flora and fauna are becoming extinct at an increasing rate every moment. Zizioulas's solution to this very practical moral dilemma is cast in symbolic language because his primary perspective is essentially eschatological. The outcomes he seeks, however, are eminently pragmatic and relate to the present. He is concerned to save the water, air, and forests. His eschatological vision of God as other assures him that this is possible. In Christ "all things" *(ta panta)* have already been reconciled, everything is already saved, the bridge between uncreated and created has been effected.[56]

However, Zizioulas is arguing that there is an urgent task to be accomplished in the historical present: if the ecological crisis is to be addressed in a fundamental way, humanity has to change the way it relates to the rest of creation. He does not believe that moral dictums or mere good

---

54. Metropolitan John (Zizioulas) of Pergamon, "The Eucharist and the Kingdom of God" (Part III), trans. Elizabeth A. Theokritoff, *Sourozh* 60 (1995) 43–44.

55. An expression of Ignatius of Antioch. See John Zizioulas, "La Vision Eucharistique du Monde et L'Homme Contemporain," *Contacts* 19 (1967) 86.

56. See John Zizioulas, "The Mystery of the Church in the Orthodox Tradition," *One in Christ* 24 (1988) 301.

sense will effect the kind of primordial change needed. He believes that the profound eschatological vision of God at the heart of Christian faith is capable of sustaining a practical conversion that will bring about salvation for creation now and into the future. Zizioulas's argument is that the culture created through the living ethos of a vibrant Christian community, centered on the Eucharist, can provide for this most powerfully. He believes that the liturgy is the key formative source of initiation into a way of living and relating to God as *other*—as a trinity of Persons—that can shape and transform humanity's relationships and behavior toward every other entity.[57]

From a very different approach, Johnson's Trinitarian theology offers an equally compelling vision of God that overcomes the hierarchical dualism of modernity. The essential elements of her Trinitarian theology address this root cause by offering a constructive alternative that is firmly founded on the tradition and "our own ancient culture." Her vision of God, however, is situated within a contemporary cosmology, "God's good creation."[58] Her feminist method, which begins from the experience of women, insists that for Christian theology to be authentic, it must both recognize and contribute to the full humanity of women. Her choice to begin a theology of the triune God with the Creator Spirit at work in the cosmos underlines the sacredness of matter, of women and women's bodies, and of all that is created. Her substantive drawing from the Wisdom tradition reinforces this sense of divine engagement with all creation; her use of female names for the one God and for the three divine Persons explicitly values women as *imago Dei* ("image of God"); her insistence that "fellowship, community, *koinonia* is the primordial design of existence" and mirrors the being of God's very self links the insights of modern science with the fruits of a contemporary Trinitarian theology.[59] All these elements, separately and together, work to deconstruct hierarchical dualism and to build a new life-sustaining vision that is based on the doctrines of the triune mystery.

As Johnson states and restates, the symbol of God does indeed function. Her retrieval of the triune symbol provides a strong unifying vision that is capable of sustaining a transforming ecological ethic and praxis.[60]

57. See Metropolitan John Pergamon (Zizioulas), "Orthodoxy and Ecological Problems: A Theological Approach," *The Environment and Religious Education: Presentations and Reports, Summer Session on Halki 1994,* ed. Deuteron Tarasios (Militos Editions, 1997) 30.

58. Elizabeth A. Johnson, "Turn to the Heavens and the Earth," 1.

59. Johnson, *Women, Earth and Creator Spirit,* 44.

60. This is also argued by Michele Anne Grimbaldeston, *Sophia Renewing Earth: An Ecological Re-Reading of Elizabeth Johnson's Feminist Trinitarian Reconstruction of the God Symbol,* an unpublished Honors thesis submitted to the School of Theology of The Flinders University of South Australia, 1997.

She is stark in her analysis that "saving the earth requires hard choices and courageous deeds in the political, social, economic, and cultural arenas. To reflect upon and promote such critical praxis, theology has the need of thought patterns that disrupt human dominance and promote the whole community of life."[61]

The vision of Holy Mystery that unfolds throughout Johnson's Trinitarian discourse does just that. It presents a way of looking at God, humanity, and creation that promotes the whole community of life. It evokes a holy space that is truly redemptive for women as well as for men and where the poorest and those who are considered "different" can be at home. It provides an image of a place where all being is valued. Moreover, this vision of divine *koinonia, SHE WHO IS: Spirit-Sophia, Jesus-Sophia, Mother-Sophia*, invites each person into relationship and collaboration with her triune self who is mercy. It redefines and reimages divine omnipotence and in so doing inexorably challenges the believer toward an ethic and to action. Divinity imaged as God-Three and God-She shatters comfortable boundaries and reminds humanity that God is the divine other and finally beyond definition. To know and to relate to this Trinitarian God has practical consequences. Relationship with such a God opens up a way of relating to the earth that is truly a "healing embrace."[62]

I began this chapter with the premise that there is a link between humanity's capacity to acknowledge God's utter otherness and humanity's capacity to respect the otherness, the integrity, of creation. I believe that the Trinitarian theologies of both Johnson and Zizioulas, different as they are, disclose this link. God's otherness is revealed as an otherness of staggering relationality that invites humanity into relation with God and all creation and so into fullness of being. These theologies show the Trinity functioning as a living symbol that contributes to an ethos that effects transforming action. They point to how this central symbol of the Christian faith can work to facilitate the participation of believers within God's life, within the human community, and with non-human creation. They show how it provides the basis for a truly Christian ethic and praxis.

At this time of "taking stock" at the beginning of the new millennium, there is a challenge to recover the delicate ecology of the human soul along with the fragile ecosystems of the earth. I have been arguing that the retrieval of an ancient symbol of God has a crucial place within this task. If the earth is to be healed, there is a need not only to heed the earth's cry "Do not kill me!"—this plea must also be heard from God's very self.

---

61. Elizabeth A. Johnson, "Heaven and Earth Are Filled with Your Glory," *Finding God in All Things,* ed. Michael Himes and Stephen Pope (New York: Crossroad, 1996) 96.
62. See Lorna Hallahan, "Embracing Unloveliness," Chapter 6 in this volume.

Australian Seascape

Well,
I was thinking about turtles.
They're harmless.
Lot of them are dying, you know.
They're being caught up in pollution,
Strangled in trawler nets
And they're swallowing plastic bags
Because they look like jellyfish.

When I go to the beach
With my grandchildren
I always say,
"Save the turtles.
Pick up the plastic bags."

You know, if they go
They are gone forever.
So why should they die?

Bonnie Hagen

Chapter 6

Embracing Unloveliness

Exploring Theology from the Dungheap

Lorna Hallahan

Life is gorgeous. It is more than pretty, it is also wild and barren. At times it has a strange and savage beauty. Discerning our belonging to the earth, we marvel. This is the mystical road described by Teilhard de Chardin that is "to become one with all, through a peak of intensity arrived at by what is incommunicable in each element. This procedure leads to a . . . unification of the elements within a common focus, the specific effect of love."[1] Western mystical traditions interpret these deep inner experiences as encounters with a holy other. Tony Kelly speaks of "a participation in the creative knowledge and love of God, the source and goal of the universe of particular realities."[2]

And yet, doesn't mystical connection within life inevitably embroil us in pain and death in ways that stir fears and the desire to flee? Sara Maitland argues that at the very point where we begin to explore what our connections within all life mean, we become estranged. She goes on to say, "Fear has estranged us from our world, which is God's world. We Christians have been warned off our territory. We have been scared to the point that we do not dare to lift a stone and look at what is underneath."[3] This chapter looks under some of the stones that are so much of our territory. Drawing on insights from my personal experience of impairment and professional life as a disability advocate, I question our responses to all

1. Pierre Teilhard de Chardin, *Toward the Future*, trans. René Hague (New York: Harcourt Brace Jovanovich, 1975) 209ff.

2. Tony Kelly, *An Expanding Theology: Faith in a World of Connections* (Newtown, N.S.W.: E. J. Dwyer, 1993) 47.

3. Sara Maitland, *A Big Enough God: Artful Theology* (London: Mowbray, 1995) 190.

that is deemed unlovely. Is our desire to cure the world of suffering found in love? Or do we act out of the dread of affliction and fear of estrangement from God, each other, and gorgeous life? Can we really help in the work of ecological healing, or do we just want to get rid of the dungheap?[4]

In her exploration of theology for skeptics, Dorothee Soelle suggests that the fear of fear stifles our pursuit of wholeness, because repression takes all our energy. She says that our natural reaction to "self-made chaos is fear and the desire to suppress."[5] Our attempts at repair are often further efforts in repression. She argues that "the holistic devotion of the new age does not articulate that every yoke will be broken. It takes its wholeness from nature and not from the humbled and offended neighbor made in the image of God."[6] Taking up this challenge at the heart of Christian faith— that every yoke will be broken—I suggest that reflections from the worlds of disability can help us to see that in the embrace of all creatures of the dungheap lies the possibility for healing. Theology's place is not to propose solutions but to speak into our fears in a way that provides hope through effrontery. By sketching ecoadvocacy within Miroslav Volf's metaphor of embrace, this chapter explores theology from the dungheap—a theology that seeks posterity for the weak through the enduring power of covenant.

Archaic Dread and the Motivations for Repair

The realization that life brings with it unloveliness, worthy apparently only of our apprehension and revulsion, may make us suspicious of the seductive loveliness of vigor and beauty. We know that life is able to discard those who are weak through accidents of conception and birth.[7] Is *this* nature that places some on the dungheap worthy of our marveling? Paul Ricoeur says that when we experience feelings that are of the order

4. The "dungheap" calls up the place of extreme degradation where Job retreats with his hideous sores. It is the place of disposal of that which is undesirable. It leaves us very uncomfortable because it reminds us of the unavoidability of waste in our bodies, in the lives we lead, in the social relations we share, and in all our processes of production. The dungheap can also take on the value of compost, of decay that releases and nurtures ongoing life. While this has very worthy ecological connections, it is a dangerous link to draw when looking at the lives of people who experience disability, for it too quickly offers a utilitarian justification for suffering.

5. Dorothee Soelle, *Theology for Skeptics: Reflections on God,* trans. Joyce L. Irwin (Minneapolis: Fortress Press, 1995) 117.

6. Ibid., 120.

7. Sometimes such accidents of conception and birth have inexplicable etiology. We know, however, that war, polluted environments, maternal impoverishment and ill health, and poor medical attention in the first years of life are the major contributions to childhood disability in all its forms.

of primal or archaic dread, we "enter into the ethical world through fear and not love."[8] In this first section I will look at our archaic dread of affliction. I will then seek to understand why, moved by these complex fears, we intervene to repair.

Writing of suffering and God, Leslie Kingsbury tells this story of his visit during the late 1940s to a home for physically and mentally deficient children.

> Some of them were frankly shocking to look at. It is doubtful whether many people ever see these crippled little folk. There they lay, some afflicted with sores, others with twisted limbs, others with grotesque, enlarged heads. Humanly speaking—as we say—there was little hope for any of them. Occasionally a little child would respond to the care of its nurses and the virulent disease would yield its grip, but the twisted arm or neck, or the misshapen head would remain, a permanent and terrible reminder of the human tragedy. As the apparent finality of this tragedy thus expressed itself, the matron observed: "There are those who would say, 'What is the use of all this?' But the gas chamber is no answer. These little cases are mostly hopeless, but they are interesting and provide valuable material for research."
>
> And yet as the priest stood and watched one little fellow, twisted and disfigured, monotonously and rhythmically moving his head to and fro, his eyes closed, his mouth half open, and realised that he had been doing that for a very long time, he was overcome with fear that the child might be suffering intensely. It is a fearful possibility. We can no more deny the possibility of this than we can affirm it to be so in any one instance. It is one of these manifestations before which the silence of the heart is disturbed by unspeakable questions and dreadful fears for humanity.[9]

Reading this, we might be struck by the honesty and the intensity of Kingsbury's reactions. He records his reactions in language that would be found too graphic and perhaps offensive in contemporary discussions about disability. Clearly, however, he does not record all possible reactions. Beauty is indeed in the eye of the beholder, and some of us might react differently in this encounter. Impairment is not of its very nature horrifying. Yet all of us will have known a time when, encountering suffering, we are immobilized and silenced. Faced with affliction, we dread the possibility of such loss in ourselves and in those near us. Certain individuals and groups

---

8. Paul Ricoeur, *The Symbolism of Evil,* trans. Emerson Buchanan (Boston: Beacon Press, 1969) 30.

9. Leslie Kingsbury, *The Way of Suffering: Essays on the Christian Significance of Suffering* (London: Longmans, 1948) 10–11. Given the contemporary commitment to promote positive images of persons with impairments, it is highly unlikely that such a graphic depiction of the conditions of these people would be recorded. In using this quotation, I acknowledge its potential to offend because it is so graphic and perhaps emotionally honest.

of people have the power to threaten our comfort and alter our lives—those with disability, those in the decay of age, those who drink or drug themselves to death, and those who starve. We cannot shield ourselves from the bodies of others or our own bodies. No matter how healthy our lifestyle, how spiritual, how ethical, how attentive to the will of God we become, we might still become seriously ill, or growing old will return us to various states of dependency on the vigor of others.

Simone Weil, writing in France during the Second World War, developed an understanding of affliction that is useful here because it combines extreme pain and social degradation with unspoken suffering—the laceration of the soul.[10] While we respond joyfully to the *loveliness* of life, *unloveliness* contains these elements of affliction: Unloveliness looks awful. It also conveys weakness and grief. It dumbfounds us and raises dreadful fears for humanity. It conveys unlove and the absence of God. This is the horrible absence of God felt by Job on his dungheap.[11] And far from being uplifted, we are brought down by it. There is no marveling at the work of a beautiful, vigorous, and creative God here. If children born to a short and painful life of overwhelming disability make a compelling case against God, so surely do all creatures of the dungheap.

The connection with ecological destruction becomes obvious here. Our context is indeed distressing. We live with unsustainable maldevelopment and global inequality in resource use, which displace and impoverish indigenous peoples, nomads, forest dwellers, and peasant peoples. We look at befouled landscapes, denuded forests, polluted streams, oceans and deserts filled with rubbish. We record widespread extinction of species. We are brought face to face with what Kelly calls the full given reality of life at this time, combining death, decay, and natural disaster with disease, plague, war, and exploitation. He reminds us poignantly that "everything is not beautiful and life has piercing tragedies."[12]

The movement of people who live with disability makes a very clear distinction between seeking a cure and working for healing. Kathy Black, in her work on homiletics and disability, defines seeking a cure as evidence of a dominant concern with *doing*. She explains that "cure is the elimination of at least the symptoms if not the disease itself."[13] Thus cure restores

---

10. Simone Weil, "On Human Personality," in *Simone Weil: Utopian Pessimist,* ed. David McLellan (London: Macmillan, 1989) 278.

11. The image of Job covered with sores and lying in ashes and muck remains a compelling symbol for the unavoidability of suffering and for the paucity of our attempts to justify God in the face of affliction.

12. Kelly, *An Expanding Theology,* 59.

13. Kathy Black, *A Healing Homiletic: Preaching and Disability* (Nashville: Abingdon, 1996) 51.

the object of the intervention to full function. Restoration to full function is indeed an admirable goal; it is often effective in treating illness. However, in this view the person, place, or object that does not respond to the cure risks being discarded and excluded from relationship: the incurable are left on the dungheap. Conversely, Black describes healing as the bringing of the person back into community. She says, "Healing is the breaking down of barriers allowing the being of the person to be affirmed."[14] In the section on embrace and ecoadvocacy I develop this view of intervention.

First, however, it is important to look further at our impulse to cure, that is, to eradicate the undesirable aspect of a person or place and to restore it to full function. Ricoeur points to the tenacious links between sin, defilement, and affliction. Describing what he calls ethical terror, Ricoeur suggests that "dread of the impure and rites of purification are in the background of all our feelings and all our behaviour relating to fault."[15] Black illustrates these links with a vivid tale:

> In recent years, a man who was born with severe physical deformities was looked upon with disgust and suspicion. Several times people said that only a pact with the devil could create such a grotesque being. They judged the man solely by his outward appearance. They found it difficult to imagine a god who would create such deformities. Therefore, they reasoned, the person with disability must be of the devil.[16]

If we find a force for evil in those with impairments, we will strive to protect ourselves from contamination. Associating fault with misfortune, we may fear retribution, even divine retribution. So we seek to cure the damage done, either naturally or by our own ravaging hands. Cure, as a means of soothing our dread, becomes evidence of overcoming the power of evil. Cure is evidence of purification, and cure is evident only when function is restored. Alternatively, even when we have dissociated much of creaturely suffering from sin, naming suffering as absurd and scandalous, we suggest that seeking a cure is compassionate and moral. Like

14. Ibid., 53. It is important to stress that the need for healing is not confined to individuals. Black talks of how "some people in the disability community feel that it would have been better if Jesus had healed the community of its common practice of ostracising all those with disabilities" (122). Betenbaugh and Proctor-Smith also stress this caution, saying that while the distinction between healing and cure is appropriate, people with permanent disabilities may find in the suggestion that reconciliation with God brings about healing an implication that their disability is a result of separation from God. See H. Betenbaugh and M. Proctor-Smith, "Disabling the Lie: Prayers of Truth and Transformation," in *Human Disability and the Service of God: Reassessing Religious Practice,* ed. Nancy L. Eiesland and Don E. Saliers (Nashville: Abingdon, 1998) 282.

15. Ricouer, *The Symbolism of Evil,* 25.

16. Black, *A Healing Homiletic,* 21.

Matron, we become determined to rescue something of use. Therefore, aspiring to cure includes a variety of motivations to repair that retain a primary concern with function. We might act to eradicate a problem, arguing that we should prevent such things from happening in the future. This contributes to a wider management approach concerned with putting things back in their places and getting them going again.

In all these motivations we act out of our rejection of unloveliness.[17] Either for reasons of aesthetic revulsion or inability to bear the pain, *we seek to repair in order to enter into relationship*. Yet it seems that at times our efforts to cure, motivated by our need to silence the voice of despair, simply expose the depth of angst and its implacability in spite of all our work.[18] Sharon Welch argues, in *A Feminist Ethic of Risk,* that to greet unsolved or inappropriately solved problems with defeat is to succumb to an ideology of cultured despair, in which we know the extent of ecoinjustice and feel powerless to act in defiance of it.[19] Cultured despair makes us yearn for the pain to stop. And if the cure doesn't work, we throw some things away. For the incurables, the dungheap becomes their final destination.

Furthermore, the dungheap, constructed by natural mishap and compounded by unjust human action, diminishes our experiences of God as powerful, loving Creator and Savior. Thrown out of our rapture with gorgeous life and stripped of comforting notions of God's power to intervene and fix things up—to miraculously prevent or cure affliction—we respond with repression, condemnation, or despair.

Yet this is not the full picture. What makes it possible for us to move beyond seeking a cure? As Black asks, "Will the community be one that brings healing through acceptance, support and encouragement, or will the

---

17. I am mindful that in using a word like "all" I might fall into the trap identified by Stephen Jay Gould. He says that "the world is a complex place. In our struggles to simplify and understand, we often identify some bugbear and then make it responsible for all evils." See Stephen Jay Gould, "Utopia, Unlimited," in *An Urchin in the Storm: Essays About Books and Ideas* (London: Collins Harvill, 1988) 218. I am not attempting to explain all evils. Indeed, it is possible that, at a superficial level anyway, repair motivated thus is beneficial for the recipient. However, I am admitting that revulsion and discomfort in the place of unloveliness are a pervasive emotion in my own responses. Furthermore, I have experienced it in my own excursion into the world of prosthetics (a process that makes up for deficiency). After many months away from home and school as a resident in a rehabilitation institution, I was "rewarded" with an extremely heavy wooden leg. It did not enhance my mobility in any way, and I was rendered a very clumsy two-legged person, when I am really an agile one-legged person. I was always encouraged to wear it when meeting the amputating orthopedic surgeon so that he did not feel so bad at seeing me mutilated by his hand.

18. In *On Human Personality* Weil talks of the inadequacy of rights language when dealing with affliction.

19. Sharon D. Welch, *A Feminist Ethic of Risk* (Minneapolis: Fortress Press, 1990) 104.

community of faith establish boundaries to protect itself from those considered unclean or cursed?"[20] Is it possible that a different starting point leads us away from actions we may construe as worthy and efficacious and into a new understanding of ethical action? Is it possible that combining our awe at lovely creation with our commitment to be open-eyed and open-armed about unloveliness could give theology a renewed ecological purpose?

Theology grounded in the distressing context of ecoinjustice can subvert managed solutions that offer potentially brutalizing remedies, patch-ups, and cleanups.[21] For Welch, the ability to hope in the face of continued apprehension is grounded in theology that "values the finite and our attempts to establish justice, and offers alternative ethical sensibilities and radically different religious imagination."[22] A theology from the dungheap *can* offer us this vision, hope, and imagination instead of a deeper journey into cultured despair. Taking up the challenge that every yoke shall be broken, we can seek another face of God. We can seek an incarnate God who is present in our work of healing.

At the outset of the Ecotheology Project that has brought this book together, we met with people who were working to heal the earth. They all spoke of a deep and moving connection to the earth, including those parts seemingly damaged forever; they expressed a desire to know more, as well as a powerful motivation to respond with careful, quiet, local nurture of desertified, deforested, and overly exploited land. Upon reflection, it seems that these activists were not acting from fear but from hope and courage. One participant, looking back over many years as a hydrologist, said, "I feel that my faith and my work are coming together."

Theology from the Dungheap

Can joining our faith and our work assist us all in the challenge of healing? Added to our awareness that we too are of nature, with all its uncertainty,

---

20. Black, *A Healing Homiletic,* 53.

21. During the 1950s the British Government conducted nuclear weapons testing in "uninhabited and empty deserts" in central Australia. Alarming numbers of the soldiers involved have suffered serious health effects. The contaminated area was left for dead. At the time the local indigenous people were supposedly moved off their land into fringe settlements. However, over the last forty years many stories have been told about who was around at the time and their experiences. The community leaders have negotiated with the Australian government for the land to be cleaned up and restored to the local people. In early 2000 the land was declared clean by the Australian government. However, indigenous leaders, environmental activists, and some scientists who were working on the cleanup operations have expressed grave concerns about the methods used and consider the area unsafe for occupation.

22. Welch, *A Feminist Ethic of Risk,* 122.

is the final owning up to being deeply implicated in the violence and mess throughout our history, producing an ecoinjustice of widening scale and impact. It is not just Gothic or morbid imagination or ethical compulsion that turns us around to face unloveliness. When we flee we too become the victims of ecoinjustice, with all its implied violence, messing up, and creation of unloveliness. We would be saying that there are some things that happen to human beings and to our world that are irredeemable, forgettable, and bound for the dungheap. We too become irredeemable.

Yet, recalling those moments of deep connection within gorgeous life and the effects of love, we know that we cannot simply stand by and let the whole thing rot. Instead of pretending that all will turn out well or casting our eyes heavenward, seeking an individual, lovely resting place, we can agree that it is life before death that matters. For Sallie McFague, the sense of connectedness and oneness requires our theology to link the mystery and the mud.[23] What is needed at this point is something that does not simply wrap the dungheap up in the rhetoric of mystery and leave it at that! A too-soon, fetishistic retreat to the numinous is to conclude that God cannot protect God's creatures—it is to declare God creative but not merciful. For the dungheap interrogates not only God's creative powers but God's mercy. Lucy Larkin, drawing on ecological science and theology, argues in Chapter 8 of this volume that "reality (which must include humanity and all earth others) is relational. God is relational."[24] We cannot avoid it.

As we face unloveliness, this relational God tells us that our future lies in what we desperately want to ignore or to change, to miraculously repair, or to mystify. And then to move on. However, relationality also gives us a God whom we can love and by whom we can be loved. Soelle stresses this as ultimately hopeful, saying, "It is in the undivided love towards God that the holistic connection to life grows."[25] Soelle goes on to say that this true love of God, undivided and given as well as received, "must place itself within the reality in which we live."[26] *This* is the God who wants to be born in us.[27] In Christian theologies, incarnation conveys a particular understanding of immanence and relationship that also directs us away from cosmic mysticism or from ecoescapism. It brings us right into present reality. Within this understanding of the incarnation of Jesus, it becomes possible to amplify all other incarnation.[28] In this we can assert with Catherine

23. Sallie McFague, *The Body of God: An Ecological Theology* (Minneapolis: Fortress Press, 1993).
24. Lucy Larkin, "The Relationship Quilt."
25. Soelle, *Theology for Skeptics*, 117.
26. Ibid., 117.
27. Ibid., 126.
28. McFague, *The Body of God.*

LaCugna that "the mystery of God is indeed the immanence, the indwelling, of divine love, for which we are made as desired partners."[29]

As desired partners in divine love, we can share "the focal point of early Christian self-understanding," says Elisabeth Schüssler Fiorenza, "which was not a holy book or a cultic rite, not mystic experience and magic invocation, but a set of relationships: the experience of God's presence among one another and through one another. . . . God's new world is not within us but among us."[30] We can take this further into all communities, not simply human communities. By developing the concept of the planet as God's body, we can see that the *earth* reveals dimensions of God's communion and the activity of the life-giving Spirit. And it doesn't stop there—this knowledge brings with it the ethical imperative to do the work of healing.[31] That is precisely the connection that Black offers us: she argues that healing is primarily concerned with *being*. And it is being that binds the mystery and the mud.

The following conversation between two men—one whom illness has paralyzed and who communicates using a sip/puff control on a computer—reveals the necessity to look beyond reducing all existence to function:

> "I was angry for about three years. It ate me up. Then I stopped being angry. Now I pray to Jesus Christ."
>
> I have no right to object to this, but I do, strenuously. I tell him that I will not pray to someone who makes him suffer like this. His mischievous eyes do not lose their sense of humour. Letter by letter, more words emerge on the green screen.
>
> "I lie here; I cannot move; however, I can listen, think, pray. How is it I feel love and where is it coming from?"[32]

Concentrating primarily on being does not exclude doing but places it on a wider plane that prioritizes interventions that facilitate relationships and do not simply fix up gross physical performance. In this short and efficient communication between the man angry about a God who seems to abandon some creatures and the man changed by love, we can see the revolutionary potential of a focus on being. Seen theologically, this focus on being widens our understanding of grace. Grace is no longer tied up in an isolated event of resurrection. God is no longer aloof but is revealed as the God who embraces all God's creatures. Through grace we know the love

29. Catherine Mowry LaCugna, *God for Us: The Trinity and Christian Life* (San Francisco: HarperSanFrancisco, 1995) 367.

30. Elisabeth Schüssler Fiorenza, *In Memory of Her: A Feminist Reconstruction of Christian Origins* (London: SCM Press, 1983) 345.

31. See Chapter 3 by Denis Edwards in this volume.

32. Michael Ignatieff, *Scar Tissue* (London: Chatto and Windus, 1994) 137.

that is passionate for the unlovable, the undignified, those who cannot even speak of love. We can know a love that embraces the angry, the hostile, and the one who will not be loved. Therefore, in affliction we might find a transcendence that is transformative, not only for suffering creatures but for those of us who try hard not to back away from the truths of extreme pain, loss, and fear. As Jürgen Moltmann says, "Healing thus consists in fellowship and in sharing with and being a part of all things."[33]

It is from these understandings that we can use Volf's major metaphor of covenant and embrace.[34] Embrace provides ways to make real our incarnation as informed and active sensibility that manifests as ecoadvocacy. For Volf, the human or social side of the (new) covenant that emerges from the incarnation and death of Jesus becomes "one way of embracing each other under the conditions of enmity."[35] The covenant must take in the conditions of the dungheap. Enmity is only one of those conditions, which include physical pain, estrangement, affliction, dread, and anxious attempts at cleansing and repair.

Clearly this means that we eschew all missionizing aspects of theologies that justify the conquering of people and places in the name of a transcendent and powerful monotheism. It becomes a theology that can speak into the fear, not one that hides the fear with pride or submits to the fear with despair. Theology from the dungheap is therefore no longer a narcotic or a fetish. It becomes a theology that is the "enemy of sentimentality and of simple solutions."[36]

Theology from the dungheap speaks to the ecological crisis through its essential and persistent inability to answer the questions. Simply because we do not avert our eyes and do not promote solutions that propose getting over it and getting on with it, our theology might not babble on in the hope that words will cover up the eerie silence of an afflicted earth and its creatures. Theology can declare its legitimate place as a voice within the cultural work necessary to resist ecological exploitation and the rejection of weakness. Yet theology from the dungheap must know its limits.

Theology from the dungheap acknowledges it limits by retaining its link to hope through effrontery—a shamelessness with naming the forbid-

33. Jürgen Moltmann, "Liberate Yourself by Accepting One Another," in *Human Disability and the Service of God: Reassessing Religious Practice,* ed. Nancy L. Eiesland and Don E. Saliers (Nashville: Abingdon, 1998) 117.

34. Miroslav Volf, *Exclusion and Embrace: A Theological Exploration of Identity, Otherness and Reconciliation* (Nashville: Abingdon, 1996). Volf offers a typology of the new covenant as embrace.

35. Ibid., 156.

36. Jeannette Winterson, *Art Objects: Essays on Ecstasy and Effrontery* (London: Vintage, 1996) 184. Winterson says that this was major a characteristic of T. S. Eliot.

den. The forbidden is primarily our deep entanglement in unloveliness. So within this savage reality of the dungheap, is it possible to imagine within ourselves an authentic desire to become an instrument of transformation through our work and our lives as (Christian) ecoadvocates? Speaking of the work of artists, Jeannette Winterson says:

> The artist is an imaginer. The artist imagines the forbidden because to her it is not forbidden. . . . Her clarity of purpose protects her although it is her clarity of purpose that is most likely to irritate most people. We are not happy with obsessives, visionaries, which means in effect that we are not happy with artists. Why do we flee from feeling? Why do we celebrate those who lower us in the mire of their own making while we hound those who come to us with hands full of difficult beauty? If we could imagine ourselves out of despair? If we could imagine ourselves out of helplessness? What would happen if we could imagine in ourselves authentic desire?[37]

Ecoadvocacy takes on just this sort of artistry. Our hope and our creative capacity lie in the offer of embrace of all those creatures and places trapped in unloveliness. This hope, building on a chastened and invigorated moral imagination that shifts the focus from doing to being, is bold enough to declare its partisanship with all that is responded to as unlovely. And it becomes the source of the energy needed for an ecoadvocacy that can place creation and all creatures of the dungheap in positions of worth. For as McFague says, "A sense of oneness with the planet and all its life forms is a necessary first step, but *an informed sensibility* is the requisite second step."[38] Surely, despite the distancing power of dread, our sense of oneness can extend to all of gorgeous life, even when it manifests as afflicted. An ecoadvocacy that affirms oneness with *all* creation cannot stop at information or at sensibility.

This oneness also calls us into a third step: the action of healing. Healing relates to the restoration of the binding power of the covenant. It is not simply reconciliation to God. How mistaken to assume that all those thrust on the dungheap are at a distance from God—that would be another way of associating affliction with sin. No, healing happens because those who are not afflicted move toward all discarded creatures and places in grace and love, free from fear and the desire to dominate. Herein lies the effrontery to dominant social values and practices that theology from the dungheap proclaims.

The third part of this chapter explores the possibilities of this effrontery: active partisanship with all creatures condemned to the dungheap.

37. Ibid., 116.
38. McFague, *The Body of God*, 128.

We approach each other with our hands full of difficult beauty, developing an informed and active sensibility, taking up ecoadvocacy with authentic desire and the drama of embrace. Healing results from a desire to restore valued being to all creatures of the dungheap. By facing our dread, we move beyond solutions aimed at cured function, beyond the passivity of cultured despair, and into the risky actions of healing.

Ecoadvocacy: The Work of Embrace

By propounding embrace, Volf returns us to understanding a relationship that is prior to aesthetics and morality.[39] Therefore it speaks directly to our primal dread. Embrace grounds engagement with a respectful witness and healing because of an eternal, unconditional, and therefore indestructible covenant. Although the covenant is violated, strained, and assaulted, the worth of all other parties can never be finally eradicated. The covenant makes it possible to strive for healing, not to bring the other up to scratch for relationship. It recognizes that nobody or nothing is outside the covenant. No condition or act excludes the other from embrace.

Volf suggests that in accepting embrace, we give away the path of separation based on a fixed identity in favor of "differentiation to describe the creative activity of *separating and binding* that results in patterns of interdependence."[40] The renewal of interdependence within the covenant becomes the work of healing for justice. It is the work of rebinding disjoined relationships and safeguarding and strengthening fragile bonds.[41] Justice, approached through a process of separation and binding, of working within and because of embrace, responding out of partisanship with the creatures and places of the dungheap, is sought through the following steps, which are linked to the four stages of the drama of embrace outlined by Volf.[42]

---

39. Martin Luther expressed it this way: "Sinners are beautiful because they are loved. They are not loved because they are beautiful." Cited in Moltmann, *Liberate Yourself,* 118.

40. Volf, *Exclusion and Embrace,* 65.

41. For a concise explanation of the complexity of seeking interdependence and resisting fusion, see Chapter 8 in this volume, where Lucy Larkin talks of the relationship swamping that happens to some women, thus compromising their separateness.

42. Volf, *Exclusion and Embrace,* 140–147.

The Drama of the Healing Embrace

Stage One: *Opening the Arms*

Volf describes turning to the other and opening the arms as the expression of desire and the capacity to create space within the self for the other. This necessary first step in ecoadvocacy becomes contemplation of our world. No longer are we averting our eyes. Here our faith includes both wonder and desolation. Volf quotes from Bonhoeffer's suggestion from prison that "faith enables us to take distance from our own immediacy and to take into ourselves the tension-filled polyphony of life, instead of pressing life into a single dimension."[43] As ecoadvocates, we fill our hands with difficult beauty because we approach our work without fleeing from feelings of awe and dread. Reflection and renewal, disclosing life in many dimensions, grow as a time of self-critique and a time of dreaming. But ultimately it is only part of the creative engine of our work.

Stage Two: *Waiting*

This very immersion, our desire for a creative and valuing engagement with the whole world, including the dungheap, our very presence, which says that we will not tolerate social processes that discard others and dominate ecosystems, could be our source of imagination. But we cannot be hasty. Volf sees this stage of waiting as initiated movement to the other, not to violate integrity but to respect the other's boundaries and desire to be left alone.

Yet it is here that we really begin to heed, almost viscerally, the lament of affliction. We wait because we cannot impose closeness on all those who have been betrayed and swept aside. We cannot ask them to gleefully respond to our offer for reconciliation.[44] Those of us who know the pain— those declared the unlovely, those abandoned on the dungheap—become wary, perched for further injury or for flight. It is here that we face what Volf has called "the affliction of memory," the memories of abuse, neglect, and past injustice that add to the present sufferings.[45]

43. Dietrich Bonhoeffer, *Letters and Papers from Prison,* trans. Reginald Fuller (London: SCM Press, 1966) 209.

44. The waiting for trust to emerge might be the end point of the drama. Yes, it may be unpalatable to ask the oppressed to return the embrace offered by the oppressor because the possibility of assimilation and abuse happens again. The cycle is well documented in domestic violence literature. This is what Volf calls accepting the "underdetermination of the outcome," in which the only element that might complete the embrace is the allure of the embrace itself. Volf, *Exclusion and Embrace,* 147.

45. Ibid., 131.

However, that is not the only history. Its antithesis becomes our grounds for connection. We seek access to the experience of past dissent. In the attempt to retrieve and reconstruct a durable understanding of this relationship we share with unloveliness in all its forms, we can assertively pursue our histories of partisanship and ecoadvocacy. The tendency to corruption of "repairing" relationships reminds us that we cannot simply go back to earlier definitions or understandings of ecoadvocacy.

However, Welch has demonstrated how memories of oppression and defeat can fuel a dangerous critique of existing institutions and ideologies that blurs the recognition and denunciation of injustice. She describes dangerous memories as funds for a community's sense of dignity: they inspire and empower those who challenge oppression. Dangerous memories are a people's history of resistance and struggle, of dignity and transcendence in the face of oppression. Dangerous memories are stories of defeat and victory, a casting of the past in terms of a present of joy, hope, and struggle. Being bearers of a dangerous memory, we link imagination to memory.[46] Dangerous memories can give us access to possible interventions aimed at embrace. They can explode what Robert Schreiter has called "the narrative of the lie."[47] The narrative of the lie justifies the dungheap. Our faith-filled effrontery aimed at embrace challenges all parts of the narrative of the lie. Resistance through our history of lament, struggle, and occasional victory enables us to respond when the waiting goes on for a long time.

Stage Three: *Closing the Arms*

Perhaps Volf's goal of reciprocity will never be met. How can the child in the nursing home, rocking in his unarticulated pain, indicate a desire for mutual embrace to the minister standing by, lost in his foreboding about humanity's incapacity to salve pain? How can we know when the afflicted person can begin to trust again? How can we know that our embrace of displaced indigenous persons does not repeat the savage history of colonization and misguided repair? How can we bring life back into a covenant with all creatures? When will oppressors no longer trick us into false reconciliation and hope?

46. Welch, *A Feminist Ethic of Risk,* 154–155.
47. Robert J. Schreiter, *Reconciliation: Mission and Ministry in a Changing Social Order* (Maryknoll, N.Y.: Orbis Books, 1992) 34. Regarding the narrative of the lie, Schreiter says that violence tries to destroy the narratives that sustain people's identities and substitutes narratives of its own. These might be called the narratives of the lie, precisely because they are intended to negate the truth of people's own narratives. The negation is intended not only to destroy the narrative of the victim but to pave the way for the oppressor's narrative.

Are we therefore condemned to passive apprehension, unable to act? No, failing to close the arms is to leave the other party exposed and essentially unsafe—invited into relationship yet ultimately recoiled from. Volf says that in embrace the host is a guest and the guest is a host.[48] McFague's reference to a sense of oneness, reinforced in opening the arms, reminds us that we are indeed the guests of our global environments, the guests of each other. And it is here that we begin to make sense of our mystic connections. Embrace within the ecological crisis and as part of ecoadvocacy starts (and perhaps ends) with compassion. Compassion can be viewed as the union of the mystic sense with action for justice. Compassion then takes in what Volf, working with Gurevitch, calls "the ability-not-to-understand." This ability-not-to-understand respects the boundaries of both parties, not suppressing and negating the other in a bear hug nor abnegating the self.[49] Furthermore, it makes it possible to embrace the victims of unloveliness without embracing the exploitation and injustice that deepens and may create their affliction, without giving more power to the narrative of the lie.[50]

We strip the narrative of the lie of its persuasive power by a combination of great uncertainty and a deep conviction that all life has ultimate significance. As Welch says: "The courage to act and think within an uncertain framework is not easily achieved. It may be that this is what is meant by faith. . . . it is a stance of being, an acceptance of risk and openness, an affirmation of both the importance of human life (its dimension of ultimate significance) and the refusal to collapse that ultimacy into a static given."[51] Surely, from an ecoadvocacy perspective, all life can have this ultimate significance, not collapsed into a static value or lack of value. The closing of the arms may work as Helen Betenbaugh and Marjorie Proctor-Smith have described: to create "a safe, hospitable and supportive space, both literally in the sense of physical and sensory accessibility and spiritually in the sense of changes of attitudes."[52] It may mean, theologically, that the touch becomes as light as the silent bearing of witness to suffering. It may be a response of assertion and struggle. It may also become powerfully protective, a forcible defense of life of the other even in the face of their abject silence and lack of response.

---

48. Volf, *Exclusion and Embrace,* 143.

49. Z. D. Gurevitch, "The Power of Not Understanding: The Meeting of Conflicting Identities," *The Journal of Applied Behavioural Science* 25 (1989) 163.

50. Schreiter's concept of the misuse of history to build up the narrative of the lie connects to Volf's work in another part of *Exclusion and Embrace,* where he argues persuasively for clear judgment of evil. Exclusion thus becomes that which breaks what is bonded and assimilates that which is separate. See Volf, *Exclusion and Embrace,* 68.

51. Sharon D. Welch, *Communities of Resistance and Solidarity: A Feminist Theology of Liberation* (Maryknoll, N.Y.: Orbis Books, 1985) 78.

52. Betenbaugh and M. Proctor-Smith, "Disabling the Lie," 292.

Stage Four: *Opening the Arms, Again*

For Volf, the release of the embrace is not the dissolution of the covenant relationship but the recognition that the parties remain differentiated, not submerged into a "we" that takes on the characteristics of the more powerful party. Neither is this embrace the final embrace; it is conceivable that the release begins the cycle of embrace over again. Within ecoadvocacy this is an essential step as it reorients us to a process of reflection and contemplation and allows us the freedom to critically assess our own active embrace, mindful always of our capacities to harm and be hurt as well as to heal. LaCugna sees this contemplation "as the active enquiry into what is, allowing oneself to be shaped by another's reality, perceiving what something or someone else is in its/his/her own right, apart from our needs and desires that another be this way or that way."[53] Thus contemplation ceases to be merely escapist and becomes transformative.

We cannot pretend that we have found a once-and-for-all solution. Alberto Melucci reminds us that social movements "are a sign of this awesome power we have over ourselves—and our enormous obligation to exercise this power responsibly."[54] The possible costs are great. We may do further harm. We may seek to assimilate rather than to allow creation to continue through differentiation. Our mistrust may leave us forever victim, not gaining from the potential of the new covenant. Conversely, we may risk annihilation by a compelling alliance with the afflicted, or, at a less threatening level, we might lose a sense of individual freedom through developing a much stronger sense of being dependent within nature.[55]

Conclusion: The Glory of God Permeates Everything

Volf's metaphor of embrace shifts us from a possibly illusory fusion into a compelling responsibility, from cosmic mysticism to compassion. Thus understood within Christian theology, embrace goes beyond the numinous feelings of connection with a God who delights in creation.[56] It contains the readjustment of complementary identities, the repairing of the covenant even by those who have not broken it, and the refusal to let the

53. LaCugna, *God for Us,* 366.

54. Alberto Melucci, *Nomads of the Present: Social Movements and Individual Needs in Contemporary Society* (Philadelphia: Temple University Press, 1989).

55. Relatedly, Weil, in *On Human Personality,* argues that at times to stand in the place of the afflicted is to be annihilated as well. This adds to the risks of proffering embrace, identified by Volf: the risks of being misunderstood, despised, or violated. See Volf, *Exclusion and Embrace,* 147.

56. The new covenant does not replace but supplements the covenant of the Hebrew Scriptures. See Volf, *Exclusion and Embrace,* 156.

covenant ever be undone.[57] Surely this is a call to see ourselves collectively, as connected by and through the life-giving Spirit. Embrace contains both connection and commitment; it takes us via an understanding of incarnation as God among us, past the fear-shattered oneness, into an informed sensibility and active ecoadvocacy. Insights about embrace can ground a claim to a meaning that affirms rather than threatens our place in humanity and all the natural world. Knowing all this, can the drama of embrace lived through our ecoadvocacy and partisanship with the unlovely eradicate the fears of destruction that so mar our respect for life?

No, and it never will. We will never attain a state when we are free from the horror and desolation of the dungheap. But will we allow our reactions to be driven only by the fear? Theology from the dungheap calls for a conscious, informed decision to speak into the fear, not to give into it.[58] We can take inspiration from Alice Walker, Afro-American civil rights activist and feminist, who, in her essay "Only Justice Can Stop a Curse," argues that we are moved to moral action by love and hope, not by guilt and duty. She says, "Life is better than death, I believe, if only because it is less boring, and because it has fresh peaches in it."[59] Or as Moltmann puts it, "Beauty is the transfiguration of life through love."[60]

Does the drama of embrace ensure that our work of healing is more successful? Once again the answer is no. The embrace precedes all other action; only then does the work of healing start. Yet there are some conditions within the dungheap that are irremediable. But these are conditions that do not call for abandonment, domination, hatred, elimination, exploitation, and indifference.[61] The dungheap does not need to signify eternal damnation. Perhaps the recognition of the embrace is all that can happen. Our theology becomes a means of responsibility and accountability: "I was implicated in the mess, I need to be implicated in the healing." Marc Ellis says that we will end up having to live the dissonance. We cannot be tempted to explain things away. Yet, he asserts, "The religious

---

57. Ibid.

58. Weil suggests that "there are certain words which possess . . . a virtue that illumines and lifts up toward the good. . . . God and truth are such words; also justice, love and good . . . to use them legitimately one must avoid referring them to anything humanly conceivable and at the same time one must associate with them ideas and actions which are derived solely and directly from the light which they shed. Otherwise everyone quickly recognises them for lies." Cited in McLellan, *Simone Weil,* 288.

59. Alice Walker, "Only Justice Can Stop a Curse," in *In Search of Our Mother's Gardens: Womanist Prose* (London: Women's Press, 1983).

60. Moltmann, *Liberating Yourself,* 119.

61. Some of the features in the anatomy of exclusion listed by Volf in *Exclusion and Embrace,* 330.

becomes a sense of the fullness of claim and responsibility to each other in the everyday events of life."[62]

Does all this mean that our sense of the glory and gorgeousness of life is lost? That all we can do is hover on the edge of the embrace of unloveliness? Is the context for theological work always the dungheap?

Once again the answer is no. Incarnation takes in all of life, and we know that life can be startlingly beautiful. Indeed, as Moltmann argues, it is this very combination of God's covenant of love and incarnation that shows forth a beauty that displaces the standards of glossy fashion and nature magazines. He says:

> Human love yearns for beauty and flees ugliness. But God's love makes righteous ones of sinners and beautiful people of the unattractive. It is because we have been the beloved of God since all eternity that we can love ourselves and consider ourselves good and true and also beautiful, and that we can like our appearance. Every one of us is a reflection of God in this world. There are so many people with disabilities, so many sick and disfigured people on whose countenance and in whose visages the beauty of divine grace is reflected.[63]

LaCugna also develops this concept of transfiguration, saying: "The Glory of the Lord permeates everything, if we could but see it clearly . . . glory is the finger of God touching everything and everyone, even when we ourselves might recoil." LaCugna looks at Jesus as the one who grasped this clearly:

> He was unafraid to touch the leper, to touch the man born blind, to touch the dead, to touch and be touched by women, tax collectors and even his enemy, the soldier who had come to arrest and whose ear Peter had gashed with his sword. In fact Jesus identified in his ministry with an odd assortment of those considered unglamourous and repulsive outcasts. And yet if we ourselves could see with the eyes of glory, we too would see God present in them . . . where we might least expect it, or at least wish it to be, God's face is recognized, God's name is called upon, God's voice is heard. And the very same people who are supposed to be without spiritual insight or knowledge of God or proper religion see with God's eyes . . . they behold in the face of Jesus the face of God. They see the face of God in him and live.[64]

Perhaps this transfiguration offers meaning to an obstinate, tenacious resurrection. We begin to see the point of our striving to know a triune God

---

62. Marc H. Ellis, *Unholy Alliance: Religion and Atrocity in Our Time* (Minneapolis: Fortress Press, 1997) 155.

63. Moltmann, *Liberating Yourself,* 118.

64. LaCugna, *God for Us,* 347.

who doesn't rescue us from our foul-ups through one grand act but through the spirit of grace that can join in all our attempts at steadfastness. This continuity empowers a healing embrace that accepts accountability, that includes grief and rage, and that stresses moral discernment. This healing embrace puts justice on the near horizon. It becomes the process, the goal, the fuel, the responsibility, the cause for celebration, and the great risk of our collective ecoadvocacy, the great risk of our *koinonia* and our hope from within the dungheap.

Embrace establishes compassion, yet recognizes the limits of our power and the inconclusiveness of our theological formulations. In this way it moves beyond the pragmatic and becomes part of the depth of mysticism, freed from the threat of fetishism. It puts culture in its place, and knowing that there are some things we cannot change, we begin to reflect morally, consciously, self-critically. The courage flowing from humility that the embrace fuels within is the very opposite of heroism. It links the mystery and the mud. And maybe, just maybe, hope will come alive. That is how we do theology in the conditions of the dungheap.

Dragonflies

Beauty is what inspires me. I love beauty. When you have had a hard childhood, you grow up looking for beauty. I have three cultures in me: Austrian, Torres Strait Island, and Aboriginal from Queensland. I love gardens, I love the sea, and I love the earth. The sea is peaceful and soothing, and so are gardens with butterflies. I often use dragonflies, which are a sign of change, because they come when the dry season is changing to wet; they are the calm before the storm. I use my art to speak the beauty that is inside me. It helps me draw out the pain. It's a way to healing and inner peace.

Hermy Munnich

Chapter 7

Up Close and Personal

In the End, Matter Matters [1]

Anthony Lowes

Harvesting the Earth

In the mind of Jürgen Moltman, the ecological crisis has reached nothing less than apocalyptic proportions.[2] Thomas Berry warns that we are facing the prospect of an Armageddon much more serious and immediate than the possibility of a nuclear war. What we are facing is no less than the consequences of our actual "industrial plundering" at the hands of "a retaliatory earth," which is "a faithful scribe, a faultless calculator, a superb book keeper."[3]

In the face of the very real possibility of such earth-wide retaliation, how have the more responsible and clearer thinkers and activists gone about setting matters to rights, bringing the race, bent upon folly that is planetary in scope, to its senses? It seems to me that in ecological writing

1. I must acknowledge that the inspiration for this title is derived, in part, from Paul Collins's title *God's Earth: Religion as If Matter Really Mattered* (Sydney: Dove, 1995), though I have transposed the key sufficiently to insinuate an eschatological ("in the end") rather than creationist bias, along with a revisiting of the central role played by divine and, I would argue, created persons ("up close and personal") in the eschatological transfiguration of the universe.

2. Jürgen Moltmann, *God in Creation: An Ecological Doctrine of Creation* (London: SCM Press, 1985) xi.

3. Thomas Berry, "Economics: Its Effects on the Life Systems of the World," in *Thomas Berry and the New Cosmology,* ed. Anne Lonergan and Caroline Richards (Mystic, Conn.: Twenty-Third Publications, 1987) 7–11.

there are more or less seven broadly categorized remedial approaches or strategies that have been employed to this point in time.[4]

In the course of this chapter I wish only to canvass these briefly and then to develop within the last of them a narrowly focused elaboration of eschatology and the relationship between persons, divine and human, and the transfigured universe. I believe that if this elaboration works, it will add significantly to the endeavors of secular ecologists and to those of ecotheologians, focusing chiefly upon the process of creation (as distinct from the process of redemption). It will help them define an appropriate worldview and value-set, indeed an appropriate spirituality and practical program, to bring healing to an already substantially wounded earth.

Schooling for Scandal

Firstly, ecologists employ the strategy most readily at hand for confronting humankind with the truth of its scandalous behavior in regard to nature. Amassing scientifically researched fact and making chilling forecasts about our problematic future, they employ the rhetoric of apocalyptic catastrophe to shock humankind into changing direction. Their appeal is to the race's instinct for survival, its sense of responsibility to the present generation currently in crisis zones and future generations yet to be born.[5]

The second strategy, still within the context of doomsday revelation, appeals to more than the self-interest of the race. It targets, rather, humankind's conscience in regard to something other than its collective survival. It points to the essential scandal of consumer greed and ecoirresponsibility as a fundamentally perverse relationship with the earth that we inhabit. It underlines the irrational nature of human behavior in this regard, pointing to the universal scope of such immorality. Far from being the "apex and summit" of the process of creation,[6] it stigmatizes the human race as the blight of creation.[7] Both secular thinkers and theologians roundly condemn the distorted interpretation that many Christians and pseudo-Christians

---

4. Denis Edwards, "The Integrity of Creation—Catholic Social Teaching for an Ecological Age," *1991 Turning Point* (Adelaide: Archdiocese of Adelaide, 1991). From 139–143 Edwards surveys the main lines of thought and effort contributing to a turn-around in ecodestruction. This paper is the chief but not sole source of my own categorization.

5. Collins, *God's Earth,* 39ff. Chapter 2, in fact, is wholly given over to this question of racial survival.

6. The phrase is alluded to in Edwards, "The Integrity of Creation," 144. It is used by John B. Cobb, *Is It Too Late? A Theology of Ecology* (Beverly Hills: Bruce, 1972).

7. Ibid., 1ff. In his introduction Collins makes this point powerfully, speaking of humans living in a "feed-lot world." His castigation of humankind as the earth's "major problem" comes to a head on p. 8.

have laid on Genesis 1:26-28, claiming it as a sanction for ecoarrogance in regard to the earth—fair game for wanton plundering and rashly manipulative intervention.

In addition to these two strategies, prophetic voices also promote a raft of positive considerations to persuade a rampaging race to come to conversion. Central to this third strategy are the considerations emanating from biocentrism, geocentrism, and cosmocentrism. Urged by scientists and philosophers, these approaches point up the autonomy of living things, the earth, indeed the universe and its processes and drastically reduce the significance of the human race within the overall scheme of things.[8] Data about the universe, at both micro- and macro-levels, cumulatively points to its *intrinsic* value, situating humankind as merely one component within the whole, rather than viewing the human as the universe's point and purpose. In this way an alternative philosophy is substituted for anthropocentrism, exciting a far humbler, more deeply respectful response from humans, in the hope that they will cease to view non-human creation as contained within the circle of their own story.[9]

Within the overarching line of thought of the third strategy is a fourth, urged by the work of ethicists such as Peter Singer, who have made a case for extending to at least the higher forms of animal life fundamental rights that were previously acknowledged only for human persons. The basic premise of their argument is that higher animals are capable of experiencing pain and "valuing" their own lives.[10]

What has been briefly outlined to this point might be termed an all-embracing scientific and philosophical foundational ecology.[11] Reflecting upon its main theses, ecotheology has framed its response at several levels. Basically, in what could be thought of as a fifth strategy, there is a strain of exegetical thought that attempts to correct the wrong-headedness of the reading of Genesis that virtually gives humankind total license in its

---

8. Edwards, "The Integrity of Creation," 139.

9. See Donal Dorr's first sentence in chapter 1, *Integral Spirituality* (Melbourne: CollinsDove, 1990) 12: "To walk humbly upon the Earth: this is an attitude which lies at the heart of an authentic spirituality." Tony Kelly, *An Expanding Theology* (Newtown: E. J. Dwyer, 1993), has an interesting subsection in section I, entitled "The Humble Self," 14, where he diagnoses our unselving and denaturing and the shrinkage of the human person, caused largely by our "narrow success" and our isolating pretension. See also Patricia Fox's, "God's Shattering Otherness: The Trinity and Earth's Healing," Chapter 5 in this volume, for the way in which she invokes David Tracy, invoking Emmanuel Levinas, on this point.

10. Edwards, "The Integrity of Creation," 141ff.

11. I would have preferred to use "deep ecology" in a looser sense than that in which it is normally used to designate a specific school of ecologists, e.g., Naess, Sessions, Devall and co., but I have eventually opted for this phrase to avoid confusion.

approach to creation.[12] This emended reading of Scripture contextualizes Genesis 1:26-28 within the whole tradition of biblical stewardship of the earth, which is highlighted in the Wisdom literature and brought to a head in New Testament passages such as Romans 8:22-23 and Colossians 1:15-17. According to passages such as the latter, creation is clearly "Christoform" in its drive toward eschatological fulfilment.[13] In the overall picture, humankind has no warrant for being the unbridled dominator. Though clearly privileged within the spectrum of creation, humans are nevertheless to be governed by gratitude as the recipients of overwhelming giftedness, of which material creation is a prime component. For this they are to exercise perpetual hieratic praise and delicate husbanding.

In a sixth strategy, also following the lead of science, ecotheology has moved beyond this traditional (still, though restrained, anthropocentric) view of biblical stewardship to assert with new force the independence of the universe vis-à-vis the human race. This prescinds from the central role of the humans as creation-come-to-consciousness,[14] the climax of the Creator's progressive efforts, and focuses upon the nature of creation as the artistic self-expression of the creating God. Such a view is intensified if developed within a Trinitarian framework, as I will point out below.

The central thesis of this sixth approach is that every individual, every species, and every genus is a projection in time, in diverse and creaturely fashion, of the transcendent truth and beauty of an intelligent and loving God. Each created being, as the artwork of God, has its own inherent value, quite apart from its relation to humankind.[15] Creation within this perspective is essentially sacramental and iconic, a far cry from being grist for the maw of the human mill. The non-human component of creation has to be handled with all the reverence of a communion wafer. Human behavior that violates the earth is a kind of sacrilege; considered within the perspective of the incarnation, it is an extension of the crucifixion.

While most ecotheologians remain within the ambit of the processes of creation in their search for "intrinsic" value, some, in a seventh strategy, invoke the increased value inherent in creation as a result of the process of redemption. This, however, is not simply added value in the light of Christ's

---

12. This view is stigmatized by John Paul II in *Centesimus Annus* as the "anthropological error": Edwards, "The Integrity of Creation," 148.

13. This more balanced view is admirably outlined in Denis Carroll's article, "Creation," *The New Dictionary of Theology*, ed. Joseph A. Komonchak, Mary Collins, and Dermot A. Lane (Dublin: Gill and Macmillan, 1988) 247–249.

14. Denis Edwards, *Made from Stardust* (Melbourne: CollinsDove, 1992) 52ff.

15. Denis Edwards highlights this insight, drawing on the insights of St. Bonaventure, in *The God of Evolution* (New York: Paulist Press, 1999) 31.

incarnation.[16] That God should find a way to identify with the limitation and distended duration of created being is marvel enough in itself and, it would seem, difficult to exceed as a gesture of divine valuing. Yet the corollary of such incarnating is nothing less than the radical transformation of the universe as a whole. It is this eschatological outcome, even more than the first and, by comparison, essentially primitive movement of creation that establishes the universe as an ultimate reality with enduring value. Such eschatological thinking makes much of the notions of the cosmic Christ and of the pervasive action of the transfiguring Spirit.[17]

The main thrust of all such philosophizing and theologizing is the establishment of the inherent value of non-human creation as a prelude to the framing of a stern prophetic challenge to human narcissism to open itself in newfound awe and reverence for all that is "other" in creation. I want to make it clear that I wholly concur with, and want in no way to detract from, this thrust. Yet I think that there is also reason to consider, as well, a newly revisited and reformed anthropic principle. It serves as a further powerful incentive, beyond that normally adduced by ecologists and ecotheologians, for human beings to cherish, to be other-oriented toward, the non-human, or rather *non-personal*,[18] universe about them and to live within it with an even more intense sense of endearment and intimacy.

Invoking a New Anthropic Principle

The new anthropic principle is not a return to previous crudely articulated anthropocentrism.[19] If, as a principle, it first of all explores the nature of person and personal communion and attempts to situate the agency of created persons within the larger framework of divine and created personal communion, it has a radically decentered orientation. It reverses the essential destructiveness of anthropocentric thinking, employing another avenue to redirect the attention of humans away from themselves and their own

16. For a fuller and more subtly nuanced elaboration of this point, see Duncan Reid's "Enfleshing the Human: An Earth-Healing, Earth-Revealing Christology," Chapter 4 in this volume.

17. See chapters 7 and 8 in particular of Denis Edwards's *Jesus and the Cosmos* (Homebush, N.S.W.: St. Paul Publications, 1991) for the development of this perspective.

18. I make this distinction and include this further consideration, since the possibility of other intelligent and therefore personal life within such a vast universe cannot be excluded. And the very possibility of such life is yet another reason to urge a decentering of anthropocentrism.

19. Tony Kelly, *An Expanding Theology,*133, deplores the false eccentricity that would attempt an "escape from ourselves" by simply "decrying anthropocentricity." He points out instead that the "human is not a centre in which we end, but a centre from which we begin; not self-centred fixation, but the self-transcendence into the other, the more, the whole."

needs and preoccupations. It extends human persons ecstatically in the direction of the non-human, non-personal others of creation.

The new anthropic principle picks up on recent work done on the concept of person and personal communion within the Trinity. It makes a somewhat obvious, but to this point in time seemingly underplayed, application of Trinitarian insight to the human person's place in the engulfing communion furnished by the three divine Persons. And, as something of a speculative corollary, it examines the possibility of a miniscule, essentially instrumental yet significant, role for human persons in the ultimate transformation of the universe.

I would argue that this decentered anthropic principle is not a contradiction tangled and tripped by its own too-cleverness. It is rather a satisfying paradox that allows us to value unabashedly humankind as the "apex and summit" of creation, as creation-come-to-consciousness, and yet to turn on its head the present largely, if not wholly, perverse relationship between humankind and the rest of creation. And this, for the very good reason (in addition to all the very good reasons adduced by the intrinsicists) that besides the swath of destruction in the environment that humans leave behind them, what they have to come to terms with is that they are also engineering their own *personal* self-destruction. I mean by this something more immediate than the already mentioned possibility that they are laying the foundation for the future destruction of the race through the current despoliation of its milieu. The reason for this immediate threat will become evident as the following argument unfolds.

The logic behind this neoanthropic approach starts with the work of thinkers such as John Zizioulas.[20] He claims to have found in patristic theology, especially that of the Cappadocians, a theology of person that has been largely ignored in the West until some aspects of it made an appearance in modern philosophy, in particular in the works of Martin Buber and John Macmurray.[21] What Zizioulas emphasizes is the primacy of person in understanding something of the mystery of the Trinity, three Persons in one God. For him, God is not so much a unity of being or substance arising

---

20. John D. Zizioulas, *Being as Communion: Studies in Personhood and the Church* (New York: St. Vladimir's Seminary Press, 1985). Patricia Fox teases out Zizioulas's thought and also that of Elizabeth Johnson (whom I also draw upon) in much greater detail in Chapter 5 of this volume, "God's Shattering Otherness: The Trinity and Earth's Healing."

21. Martin Buber, *I and Thou*, 2nd ed., trans. Ronald Smith (Edinburgh: T&T Clark, 1958); *Between Man and Man,* trans. Ronald Smith (London: Fontana, 1961); and John Macmurray, *Persons in Relation* (London: Faber and Faber, 1961). The question of the influence, if any, that these philosophers may have had on each other's almost identical thought about the human person is fascinating, though no immediate answer seems to be forthcoming.

into a threefoldness of Person, as it is for Augustine and for most thinkers following him in the Western tradition.

God's unity is rather grounded in the Person of the Father, who gives rise to the Son and to the Spirit. It is their union that comprises the being of God. According to this view, the Person of the Father is the *aitia* ("root" or "cause") of the being of God.[22] The Person of the Father is the bearer of being, or in Greek, an *hypostasis* (that which "stands beneath"). Yet according to Zizioulas and the Cappadocian theology from which he draws his inspiration, the other two Persons are also *hypostaseis* and, though in some way dependent on the Father as *Arche* ("Source"), are fundamentally equal. Subsequent theologians who want to establish the perspective of an egalitarian model of the Trinity, especially feminist theologians, would want to extend the argument further. They conclude that if all three *hypostaseis* in God are fundamentally equal as *hypostaseis* and the generic quality of *hypostaseis* is that they are "being bearers," then they are each and all in this radical sense *aitiai* of the divine being. More of this shortly.

Persons in Communion

But this is not the first or most profound of the inversions of prevailing thought that Zizioulas engages in. Not only does Person, as *hypostasis,* take precedence over and "result" in divine being, but each Person as a distinct *hypostasis,* that is, a distinct particularity in its own right, is seen to be so constituted because it is first and foremost *not* something *in its own right.* It is, at its deepest level, something which is what it is only because of its relationship with the other Persons of the Trinity.[23] Each is something distinct because of the peculiar way in which it is networked with the other Two. When we speak of the relations between Father and Son and Spirit, we are speaking of much more than the psychological and the moral and the affective—as we do when referring to the bonds between created persons. We also understand that relations between them bind them in terms of being, that they are bound by *onto-relations.*[24]

It was left to the West's premier theologian, Thomas Aquinas, to stress that the Persons are, in fact, themselves *subsistent relations.*[25] It is in this fashion that it is possible to see that the communion between the Three is an *onto-communion.* Thus the Three, as Persons in relation, are each so *for* and *in* the

---

22. Zizioulas, *Being as Communion,* 39.

23. Ibid., 17–18.

24. For this precise term see Thomas F. Torrance, *The Christian Doctrine of God, One Being Three Persons* (Edinburgh: T&T Clark, 1996) 194 and elsewhere.

25. *Summa Theologiae,* Ia, q. 29, a. 4.

others that they cannot but comprise one *being*. The being of God is a communion of three Persons in eternal ecstasy (an eternal venturing out of themselves into the others and out of themselves to welcome the others within).

Hierarchy or Egalitarianism in Divine Communion

The second step, already signaled in the logic that leads toward the eschatological conclusion that I have in mind, derives from the insight of feminist theologians such as Patricia Wilson-Kastner and Elizabeth Johnson and Western theologians generally.[26]

Despite the avowals of Zizioulas and of the Cappadocians before him, who made much of the fundamental equality of the Persons in God, there is about their theology a residual sense of personal hierarchy in which the Person of the Father seems inevitably dominant. And especially in this function of being-bearing, or acting as the *hypostasis* or *aitia* of the being of God, indeed of the being of each other Person. It is difficult to see how the being of the Son and of the Spirit is not in some sense derivative.[27] It is, then, difficult to cast off the must of hierarchy from this model.

According to more recent theology, the communion between the Persons of the Trinity is absolutely egalitarian. In feminist authors, the ascription of absolute equality is justified through appeal to traditional Western theology, which has started from the substance or essence of God as the source of divine unity and gone on to explore personhood and persons within God largely unfettered by a sense of ontological hierarchy. For such thinkers, feminist and otherwise, the Father is as dependent for his being on the being of the Son and of the Spirit as they are on the Father. The encompassing *(perichoresis)*, the communion *(koinonia)* that they comprise is wholly reciprocal.[28]

The Dynamism of the Divine

If we take the trouble to explore the implications of Aquinas's doctrine of subsistent relations, it is clear, perhaps even clearer than in the theology of Zizioulas and the Cappadocians, that an eternal dynamism is at the heart of the communion of the Three. In his thinking, the divine rela-

---

26. Patricia Wilson-Kastner, *Faith, Feminism and Christ* (Philadelphia: Fortress, 1982) 126, and Elizabeth Johnson, *She Who Is: The Mystery of God in Feminist Theological Discourse* (New York: Crossroad, 1992), especially 216ff.

27. Wolfhart Pannenberg, *Systematic Theology,* trans. Geoffrey W. Bromiley (Edinburgh: T&T Clark, 1994) 1:280.

28. See Pannenberg's discussion on the radical equality and unity of divine Persons, *Theology,* 1:307ff., and also Johnson's *She Who Is,* especially 218ff.

tionships have nothing in which to lodge other than in themselves. In fact, there is no *lodging* at all but an eternally flowing action of three relations, which are engaged in fluid ontological *inexistence* (to use an ancient theological term), without beginning or end.

In this sense God is pure act. But such a notion does not represent God as some kind of fixed perfection exercising God's own static being. Nor is God a larger and more transcendent version of Atlas, upholding the universe (rather than the earth) in some sort of immobile stance. God is a mystery of engaged living and loving personal communion, which erupts eternally within God's own being and explodes into a fragmented creaturely universe (at least one), fraught with the force of God's own exuberance.

According to this line of thought, it is not simply the Father who is the *aitia* of the Trinity and of the being of God. Indeed, it is not each of the Persons taken individually, though each is identical with the being of God. Rather, it is the Three taken together in dynamic communion *(koinonia)* who confer and bear being. Being comes about through Persons in communion. Being *is* Persons in communion.[29]

Divine Persons and Ontological Density

Another contemporary theologian, William Hill, has coined a phrase that brings out even more starkly just what the Persons of the Trinity effect in each other's regard and in regard to the substance or essence or nature of God. In his words, divine Persons, and, as I have argued, divine Persons in communion, bear or bestow "ontological density."[30] The bestowal of such ontological density forever remains a personal action, obviously within the Trinity but also in regard to creation itself.

Recent work on the Trinity has moved beyond the doctrine of appropriation, which had been normative within the generally accepted tradition.[31] According to this doctrine, the Persons in God act as one in regard to creation. They act through the divine nature or substance, indistinguishably. But, as the doctrine goes, given the differentiation within the Trinitarian dynamic, by reason of the processions and relations of origin of the Persons, certain aspects of the creation process could be attributed to each of them as entirely appropriate—yet a sort of pious fiction.[32] It is

29. Zizioulas, *Being as Communion,* 18.

30. William Hill, *Three-Personed God* (Washington: Catholic University of America Press, 1982) 75, 77, inter al.

31. For a much more thorough discussion of this shift, see Denis Edwards, "'For Your Immortal Spirit Is in All Things': The Role of the Spirit in Creation," Chapter 3 in this volume.

32. Walter Kasper, *The God of Jesus Christ,* trans. Matthew J. O'Connell (New York: Crossroad, 1994) 282f.

now, however, commonly agreed that each of the Persons contributes to the action of creation in a *really* distinctive fashion and that this really distinctive fashion is closely linked with the relational peculiarity of each Person within the Trinity.[33]

An Icon of Inviolable Subjectivity

Without exploring in detail what this would mean for each divine Person at this point,[34] it is sufficient to note that each of the Persons would be operating as a subject, intending and willing the existence of the universe, with its vast diversity in unity. This would amount, as traditional theology understands, to a sharing of the being of God with discrete, dependent creatures. But it is no longer possible to think in the impersonal thought categories of essence operating upon actuated essence. If God's being is personal communion, what the three Persons share first of all is the divine-being-in-love, which stems from their own subjectivity.

Thus, as ecotheologians and ecophilosophers point out, each being within the cosmos should be viewed more or less as *subject*—at least insofar as it partakes of the subjectivity of the divine personal subjects who give themselves to and through it—as well as *object*. Each created being, non-personal as well as personal, is the "subject" of its own being, exercises the act of its own being, its *esse,* according to Aquinas. It "subjectively" expresses being, it *sub-sists*.[35] It does this, held in being within the free exercise of subjective creativity of each of the Persons of the Trinity. It is fathered or mothered into being, rescued from non-being, endowed with brute being by the *Arche,* the first Person of the Trinity. It is shaped in its particularity, its *this-ness,* its form, by the Word. It is located within the larger relational field or communion of being, divine and created, by the Spirit. In both these closely linked and complementary senses, the universe and its components are *subjectively* expressed objectivities.

In addition, rooted in divine and created communion of being, the universe and its components furnish iconographically or sacramentally subjective moments of encounter *between* and, for created persons, *with,* Persons of the Trinity. Created beings and created being as a whole are what the divine Persons personally and freely express. Created beings and created being as a whole are what they give to each other, along with the

33. Tony Kelly, *The Trinity of Love* (Wilmington, Del.: Michael Glazier, 1989) 131f.

34. See Chapter 4 by Duncan Reid and Chapter 3 by Denis Edwards in this volume for fuller treatments of the incarnation and the role of the Spirit in creation, respectively.

35. See *Summa Theologiae,* Ia, q. 44, a. 4, for example, where Aquinas makes it clear that whatever subsists may be considered the "subject" of its accidental qualities.

prior gift of themselves. This is the foundational meaning and purpose of non-human, non-personal created being. This ensures that created non-human, non-personal being essentially transcends the circle of the human story. Yet within this transcendent gifting, created being in its vast diversity is what the divine Three also give to human persons,[36] as love gifts to be revered because of their subjective otherness.

A Universe Held in Heart of Divine Communion

Such a view is compelling enough if we remain confined within the dichotomy of *ad intra—ad extra* thinking in regard to God's immanent and economic action. But it is doubly compelling if we keep in mind that even within traditional theology, creation is viewed, not just as a reality to which God is present *ad extra,* but also as a reality perpetually present *within* God.[37] This is, of course, not to conflate the divine and creaturely and reduce the relationship between the two to pantheism. Even as viewed within God, creation is not part of the essence of God; it is not simply an extended "face" of the divine, configured with temporal and spatial features. Creation is held *within* God, but still *distinct,* through the exercise of God's personal freedom.

And if this view of things is extended to encompass the insight that each of the Persons acts subjectively in regard to creation, it becomes apparent that the universe, and each component within it, is held within the interstices of the personally engaged divine communion. It is the Trinitarian process within God, which furnishes the "place" or "space" in which creation is subsequently to exist.[38] This is the space arising from the Father's withdrawal to allow the Son to be. It is the space maintained and bridged by the action of the Spirit, anchored within Father and Son. In this sense the Spirit broods over the articulation of Trinitarian communion, over the space within God. It is entirely consistent, then, that Genesis should depict the Spirit as brooding over the origin of creation,[39] occurring as it does within Trinitarian space.[40]

---

36. And this would be true of any other possibly existing personal creatures in the universe, created by the superabundant loving of Trinitarian personal communion.

37. Edwards, *The God of Evolution,* 28ff.

38. John O'Donnell, *The Mystery of the Triune God* (London: Sheed and Ward, 1998) 160. See also Jürgen Moltmann, *The Trinity and the Kingdom of God,* trans. Margaret Kohl (London: SCM Press, 1981) 108f., where he ascribes the derivation of this notion to Isaac Luria's concept of *zimzum.*

39. Genesis 1:2.

40. This same point is made more tellingly by Denis Edwards in his chapter on the Spirit (Chapter 3 in this volume).

Such is the understanding of creation as shaped by Trinitarian doctrine. The preciousness of the universe and each of its components is further enhanced if they are viewed formally from the additional perspective of grace, redemption, and eschatological transformation.

A Further Strategy for Discovering Cosmic Value

There are essentially two views of the incarnation. The prevailing view until more recently was that the Word became flesh as a backup or repair strategy to heal a universe seriously maimed by the fall of humankind. The other view, till recently a minor tradition, understood the incarnation as the endpoint of creation and the mechanism through which the creation would be brought to perfection through the leverage of the death-resurrection of Christ.[41] In this fashion it would be brought ultimately, with an even more intense intimacy, into the heart of the communion of the three Persons, thus eliminating time (as we know it) and mortality, though not space and materiality. In addition to the exploration of the doctrine of creation, this has been invoked by ecotheologians, as I have noted, as a further strategy for the discovery of intrinsic value in the universe.

Yet in exploring the transformation of the universe, what does not seem to have been explored with sufficient emphasis and clarity is the role of the human person in contributing to this transformation. Admittedly, the human person is exhorted, in the context of the doctrines of creation and eschatological transformation, to value the universe and work to heal and perfect it, as a respectful though privileged participant in its processes. But the burden of transformation is left to God, the Trinity, in and through the risen Christ, in and through the Spirit.

Where is humankind in this eschatological process? Clearly humans are recipients of the power of Christ's grace and therefore living within the communion of the three Persons of the Trinity in a way beyond that afforded by the mechanics of the creation process itself. As recipients of the grace of Christ, humans have been elevated to a "higher cleave of . . . being,"[42] because, if the life of grace is lived, that is, with personal reciprocity within the Trinitarian communion, to and through death, the human

41. Matthew Fox, *Original Blessing* (Santa Fe, N.M.: Bear and Co., 1983). The contrast is to some extent the theme of the whole book. See Appendix B, p. 316, for a schematized comparison.

42. See W. H. Gardner's employment of the phrase in "Introduction," *Gerard Manley Hopkins, Poems and Prose*, ed. W. H. Gardner (Harmondsworth, Middlesex: Penguin, 1953) xiii. And for at least one of its original usages, see Gerard Manley Hopkins, *Sermons and Devotional Writings*, ed. C. Devlin (London: O.U.P., 1959) 154: "In that 'cleave' of being which each of his creatures shews to God's eyes alone."

person is granted an unmediated vision of the mystery of God.[43] To enable this personal elevation, human persons undergo divinization or *theosis*.

The question is, are divinized human persons mere passive recipients of the transformation wrought by the Trinity in and through Christ? If this question is more sharply focused upon the moment of death and beyond, are human persons simply recipients of glorified personal fulfillment, or are they elevated and potentiated, freely functioning *instruments* in regard to the ultimate integrity of their transformed body-selves and, in a very limited fashion, in regard to the transformed universe that will be the eternal milieu of their body-selves?

Again there are two chief and opposed schools of thought. The one lays complete stress upon the action of the Trinity in and through Christ. It suggests that the point of the universe, vis-à-vis humankind, is simply as a moral proving ground. It is the re-creative action of the Trinity in Christ, which establishes not so much a transformed but a substitute universe beyond the end of time.[44] It is clear in this view that the engagement of human persons with the universe at large has only a transient value. At the very best, such action *magnifies* the persons so engaged; it has only a provisional benefit for the cosmos and seems to lie outside the Trinity's and Christ's ultimate transformative action.

The other view claims that the energy that humans expend in loving action for each other and in and for the world in which they exist is, within the overall action of the Trinity in Christ, itself continuous with and contributory to the transformation.[45] Among those proposing the latter view, there are two who ascribe a special efficacy to the human soul or spirit in death. Their views are drawn from a philosophical and theological understanding of the nature of the soul or spirit, especially as it is situated within the horizon of redeemed and graced reality. Both Ladislaus Boros (soul) and Karl Rahner (spirit) make much of the transcendental trajectory of the spiritual core of the human person, straining as it does beyond all limit in its created milieu. They emphasize, too, its radical orientation toward matter and its essential function as the form of human bodiliness.

These two trajectories or orientations coincide and come to a completion in the moment of death, which in the terminology of Boros is the ultimate moment of fully realized personal truth. Making use of Aristotle's

---

43. Zachary Hayes, *Visions of the Future: A Study of Christian Eschatology* (Wilmington, Del.: Michael Glazier, 1989) 196f.

44. Joseph Ratzinger, *Eschatology*, trans. Michael Waldstein and Aidan Nichols (Washington: Catholic University of America Press, 1988) 213ff.

45. Karl Rahner, *Theological Investigations* 10 (London: Darton, Longman & Todd, 1973) 270.

categories of form and matter, both thinkers ascribe an in-forming role to the soul or spirit in regard to the body—in life, but also and especially in death. Boros, for example, contemplates the theoretical threat that death poses for the soul, suggesting that its very reality is set at risk in the destruction of the body. He maintains that theoretically it is the moment of the soul's most vulnerable indigence. Yet within the context of resurrection and final universal transfiguration (also events within the ambit of the death moment), the human person (his thought is focusing on the soul in particular), in complete possession of its personality, *produces* "a corporeity of [its] own" and so is completely and freely "'self-posited,' right down to the most hidden fibres of [its] reality."[46] According to Rahner, the human spirit is responsible for *expressing* the risen body-self in death, once it has "become one with the pneuma of God and its bodily existence."[47]

If the latter school of thought is allowed to prevail, can a convincing account be tentatively elaborated as to the possible mechanism through which human beings might contribute instrumentally and partially to the ultimate form of the universe that they will eternally inhabit—perhaps utilizing insights of Rahner and Boros and placing them within the context of the theology of divine and human personhood?

Within a Framework of Trinitarian Understanding

I would argue that Trinitarian theology furnishes us with the beginnings of an understanding of the eschatological status of the human person, potentiated through the grace of Christ, which is essentially and ultimately transformative, though in a limited fashion.

To backtrack briefly. If we accept with Zizioulas that it is the personal in God that is preeminent and person is the *aitia* of divine being, in a word, its being-bearer . . . If we translate this into William Hill's conveyance of "ontological density," to reinforce the concept . . . If we listen to the voices of current feminist and non-feminist theologians stressing the absence of hierarchy within the Trinitarian communion and ascribe the function of being-bearing or the conveying of ontological density to all three Persons, in particular to all three precisely as they are in communion . . . If we trace the responsibility of the persons as subjects to their role in creation . . . If we look closely at the notion of grace and consequent *theosis* or divinization that this confers upon the human person . . . If we take cognizance of God's *modus operandi* in the creation process (where God acts

---

46. Ladislaus Boros, *Moment of Truth: Mysterium Mortis* (London: Burns & Oates, 1965) 62, but also 75f.

47. Rahner, *Theological Investigations* 2 (1969) 214.

in self-limiting and respectful love to achieve God's ends indirectly through secondary causes—created agents and creaturely processes) and make an assumption that God would be likely to continue in the same mode when effecting the transfiguration of the universe . . . And if we incorporate within this framework the body-soul/spirit integration emphasized by Boros and Rahner, especially as this is realized in the moment of death . . . *then* we are led to the conclusion that there is something about the divinized human person that approximates, because it shares in, this being-bearing, being-conveying function found in the three divine Persons.

I would further suggest within this frame of reference that human persons, caught up in the cosmic Body of Christ and therefore within the divinizing communion of the Three in God, are the freely collaborating and contributing instruments who, in this contextualized sense, contribute to their own glorification in and beyond death. More than that, since embodiment would seem possible only within a material universe, it makes much sense to suggest that the divinized human person, in communion with the Three in God and with all other divinized human persons, contributes instrumentally and necessarily, in a partial and limited and localized fashion, to the glorification of the universe.[48] In some sense human persons themselves, within the overall being and action of the divine Persons in communion and in and through the incarnate and redeeming Christ, partially confer upon it new and endless being. In this way it is possible to see that the new universe is continuous with the old and *assumes* its historicity.[49] It is not some sort of divine recloning simply brought about by an interventionist, magical, disjunctive deity. It is a universe in which humankind, within Christ, within Trinitarian communion, has appropriately become ontologically enmeshed by its loving action through time.

But clearly such an enmeshing is the very antithesis of selfish and unrestricted exploitation and despoliation. Such enmeshing is only effected through loving, ecstatic action that mimics and embodies the ecstatic dynamism of the Persons of the Trinity. The new universe is thus, within the limits described above, personally conceived and gestated, if you will. Such a conceiving occurs within a series of interlocking circles of personal communion.

The three largest and most all-embracing of these circles are centered in divine Persons and hold, within their overlapping, the communion of created being. In the manner of ecstatic ellipses, two of these radiate from incarnate Word and missioned Spirit. Within these again there are the

---

48. This could be further extended to include all other possibly existing and presumably (if they were to be created) similarly divinized persons.

49. Rahner, *Theological Investigations* 10, 260ff., especially 270 and 289.

myriad overlapping and interconnecting circles of communion of which each graced created person is the center. It is as enmeshed in this conceiving and gestating intersecting personal communion that human persons are engaged in their own small conceiving. Such conception and gestation occur within and through human persons only as they approach, empowered by the divinization of grace, the utter self-othering of divine Persons and the selfless devotion (even to the point of a sacrificial death for all human others and all created others) evident in Christ.

If the overall complex of personal communing is kept in mind when we attempt some understanding of the final eschatological transformation of all things, it is clear that such personal transformation is fundamentally Trinitarian and Christic. Invoking the new anthropic principle is not by any stretch of the imagination an attempt to reintroduce the old anthropocentrism into ecological discourse, but in mufti.

Such a conclusion is driven home with even more force if we keep in mind the features of personal communion as modeled within Trinitarian communion. Personal communion after this fashion is to be egalitarian and reciprocal. Though persons are fundamentally ecstatic and are what they are by what they surrender of themselves in favor of others, they are so within a hypostatic configuration. Each person is a node or nexus within a relational or communional field. Each is a particularity, albeit a particularity formed by its *germanitas,* its germaneness to the other(s).[50] Any dissolution of particularity, of hypostatic or enstatic reality, would nullify *personal* communion and reduce it to some sort of monistic simplicity or unvarying ontological soup.

Any communion that divine or human persons would be involved in, that would essentially be personal communion, would have to preserve this diverse particularity within overall relational unity. Hence, even and especially when relating to non-personal creation, divine and human persons would be relating (in the manner of all personal relating) to each non-personal created being as irreducibly other, disparate, diverse. This would be doubly so of human persons, since their own subjective objectivity as created beings, as created particulars, depends upon the personal expressiveness of divine Persons in fundamentally the same way as does the subjective objectivity of created non-personal beings. The enstatic or hypostatic particularity of the latter is a given, quite apart from and prior to any human communion that might encompass them.

---

50. This is a usage employed by St. Bonaventure. See Edmund J. Fortman, *The Triune God: A Historical Study of the Doctrine of the Trinity* (Grand Rapids, Mich.: Baker Book House, 1982) 34.

Reciprocity Between Non-personal and Personal Created Being

Finally, there is between human persons and created non-personal being an ineradicable reciprocity in which it is only in the exercise of its own *esse,* through the exercise of minimal subjectivity as I have previously outlined, that non-personal being affords human persons the opportunity to establish, to increase, and to preserve their own personal substance. It is only by relating to non-personal beings, as well as to other created persons, selflessly in love that human persons are persons at all.[51] It is by being such a relational milieu that non-personal created being bestows personal substance upon human persons as they, within the complex of interlocking circles of personal communion, contribute to enduring transfigured being of non-personal creation. This dependence for their own personhood on non-personal created beings is, of course, heightened by human persons' relationships with them as sacraments or icons of encounter with divine Persons (and within such symbol-izing, by human persons' relations with them as extensions of the enfleshed, embodied, incarnate Christ).[52] Finally, this dependence rests on non-personal created beings in their capacity as sacraments or icons of encounter with other human persons.

Both of the previous two considerations of reciprocal relationship and dependence are true of human persons and their non-personal counterparts in the current universe. It should also be noted *within the transformed universe* that transfiguration of the non-personal universe is as necessary to human persons transforming within the power of the resurrection of Christ in and through death, as they to it; without the framework of the "new heavens and the new earth," human body-persons could not continue to be embodied beyond death, could not therefore *be* beyond death, at all.

This line of thought offers one additional way of conceptualizing hellworthiness. It is to understand any wholly self-enclosed, self-aggregating, irresponsibly consuming, despoiling, violating human agents as severed from eternal divine communion precisely because such persons would be unable to find (in death as in life) any place in or portion of the transfigured human communion and material universe that could serve as their home. The reason for such inability? The failure to gain a foothold within

---

51. This is at once to affirm and to take Macmurray's basic insight further than Macmurray is prepared to go. In *Persons in Relation,* 24, he outlines the theme of what is a second volume on the nature of the human person. In essence it is that "the idea of an isolated agent is self-contradictory. Any agent is necessarily in relation to the Other. Apart from this essential relation he *[sic]* does not exist. But, further, the Other in this constitutive relation must be personal. Persons, therefore, are constituted by their mutual relation to one another. 'I' exist only as one element in the complex 'You and I.'"

52. See Duncan Reid's extended argument in this regard in Chapter 4 of this volume.

eschatologically transfigured being, through *a lack of selfless, loving relation*. Such an inhuman being would *be* hell, would *not* be a person or personal in any real sense at all.[53]

Conclusion

If human beings are to discover a reason to reverse the great harm
done to the earth to this point, they would do well to look to the urgent healing of their environment while it is still viably supportive of human life.
But it would seem that they would have to draw upon an inspiration, deeper
and more altruistic. They must do so by subscribing to some sort of broad-
ranging foundational ecology. And that, as I have already outlined, would
be arrived at inescapably by way of an open biocentrism, geocentrism,
cosmocentrism; by way of a critically select scientific and philosophical reflection on the nature of creatures, especially living creatures; and by way,
ideally, of theological reflection on the process of creation.

One of the questions to be addressed in this latter discourse of ecotheology is the question of the ultimate fate of the universe and its eschatological shaping. It is within this framework of reflection that the universe is
discovered in all its splendid inviolable subjective objectivity, its sacred otherness. It is enfolded in the heart of the communion of divine Persons, of
Christ *and,* within this larger and primal enfolding, to some degree within divinized humanity (and any other possible similarly divinized personal race).

The universe is "other," but it is also an otherness with which the
human person is in intimate *dialogue,* rather than a distanced, guilt-ridden
(if respectful) *dialectic*. This calls not only for reverence and remorse but
for engagement; for personal encounter with the sacred created and redeemed others, the iconic or sacramental others, through whom and
through which encounter with the creating and transforming Others of the
Trinity is facilitated.

Yet there is something even more personally and intimately *engaging*
in the final vision. Human persons (along with any other created persons)
and the universe at large are one because created persons *are* the universe
come-to-consciousness, *are* stardust made luminous by self- and other-
awareness. This is a central truth of science and revelation that cannot be
gainsaid. Human persons are one with the universe because they and the

---

53. Such a baleful conclusion would serve as the obverse side of the thought of
Zizioulas in regard to the need for the human "biological hypostasis" to become an "ecclesial hypostasis" to confront the destructiveness of death. Likewise, Martin Buber *(I and
Thou* and *Between Man and Man)* and John Macmurray *(Self as Agent* and *Persons in Relation)* would not allow the human self any personal reality at all unless it operates as a kind
of *node* or *nexus* within an actual relational field.

universe are held through the ongoing act of creation, not merely in the abstraction "in being"; they exist in a universe "located" in the secret recesses of the mystery of Trinitarian communion. Human persons and the universe are one, since together they are eschatologically "one body" in Christ through the power of the Spirit.

But even within this larger wisdom there is a smaller, perhaps in the end more humanly endearing, reason to cherish creation. I would argue that human persons, in life and especially in death, lay hold definitively of their own bodiliness, of their personal being, and are intimately related to the bodiliness of other human persons and to the bodiliness of at least a limited wider material milieu—their personal habitat, their personally engaged portion of the universe. They contribute to the ontological transfiguration of these (within the larger Trinitarian and Christic transfiguring) as they contribute to the ontological transfiguration of their own personal body-selves, to the degree that they have managed to love them unconditionally and serve them selflessly.

In the last instance, and within the horizon of the persuasive insights and arguments adduced to lead them to earth-healing and indeed cosmic healing, human persons may well act ecologically, impelled by a dawning sense of the intimacy between themselves and all created others, in whom only they themselves are selved, are personed. It could be said that they are *personed* insofar as they *in-person* others, that is, take others into their own personal substance, insofar as they dispossess themselves and dispose of themselves *for* and *in* others, including the non-personal "others" of creation.[54] By contrast, to live lovelessly within creation, destructive of creation, is to consign themselves individually and collectively to non-personhood, to invite immediate personal annulment. Their own personal integrity rests upon their preservation of the integrity of all around them, who are, which are, "flesh of [their] flesh."[55]

Their awakening to such deep belief and valuing, the most fundamental of foundational ecologies, would surely also include something like the ecstatic astonishment of Adam at the first sight of Eve. Their perception of the universe, as Adam's of Eve, would be of an astounding sacred beauty in distinct otherness, but also in breathless communion; in some sort, even, of personally extended mystical self-identity that is truly Trinitarian in its configuring.

54. Talk of *personing* and *in-personing* is more than a novel affectation. It is an attempt to give some definition to the ontological effectiveness of the potentiated human person prior to and in and through death that situates this within the relational and "communional" (to use Robert Kress's term) and Trinitarian, rather than confine such discussion to the narrow metaphysics of matter and form.

55. Genesis 2:23.

The Frill-Necked Lizard

These lizards are free,
and they are also protective.
When they run, they look so free—
they dance, they're happy and full of life.
Drawing them I feel peaceful,
like I'm relating to them as I do to the dragonfly.

Hermy Munnich

Chapter 8

The Relationship Quilt

Feminism and the Healing of Nature

Lucy Larkin

If God is thought of as present in creation by bringing forth
all things in life-giving interrelations
and by a constant process of renewing life-giving relations,
then humans are most expressive of God's presence
by manifesting this creative and healing process.
Sin breaks this creating and healing process
and creates distorted relationships of domination and exploitation
that aggrandize one side of the relationship
by inferiorizing and impoverishing the other side.
Sexism and all forms of exploitative domination are thus
not parts of the image of God, but forms of sin.
Salvation is the process of conversion from distorted relationships
and the regrounding of selves in healing relationality.[1]

In theological imagination, where does earth-healing come from? How do we "reground ourselves in healing relationality" to include non-human nature? Traditionally, the healing dynamic within theology has been enshrined in doctrines of salvation, reconciliation, redemption, and atonement. Etymologically linked to making whole, the healing is in the righting of wrongs, in bringing together some kind of rupture, or in a vision that things can be other, better, than they are. Hope is implicated.[2] God is deeply involved.

---

1. Rosemary R. Ruether, "Christian Anthropology and Gender: A Tribute to Jürgen Moltmann," in *The Future of Theology: Essays in Honour of Jürgen Moltmann,* ed. Miroslav Volf and others (Grand Rapids: Wm. B. Eerdmans, 1996) 251.

2. See Chapter 9 by Gregory Brett in this volume for a complementary working through of Christian hope.

Global environmental devastation requires an innovative configuring of this healing dynamic, one that is widened to include restoration, reparation, and remediation.[3] Metaphors of sickness, degeneration, and dysfunction, and concepts such as woundedness, collapse, and non-sustainability assist us in describing and appreciating situations where flourishing is denied to the earth.

Grace Jantzen contrasts a theology of flourishing with a theology of salvation. For her, flourishing is linked with blossoming, vigorous and luxuriant growth, and good health.[4] Salvation, on the other hand, denotes rescue and thereby places humanity in a largely passive position.[5] Jantzen argues that a theology of flourishing assumes the interconnectedness of people and ecosystems, naturally leading to justice and protest as it aims at physical health and adequate material provision for bodies.

Healing is a prerequisite for flourishing. In feminist theological discourse, healing is described as creating right relationship. What would it take for me, for any of us, to create right relationship with nature?

This chapter resembles a quilt, or at least the process of making one. The works of feminist theologians are the raw cloth. They have been selected because they share color and tone: the red of feminism, the blue of process thought, the green of ecological sensibility. Patterns vary, as does the mix of colors. For instance, Carter Heyward and Rita Brock are more red than blue, with the odd touch of green. Marjorie Suchocki is predominantly blue, Catherine Keller is red and blue with splashes of green, while Mary Grey and Sallie McFague are a fairly even combination of all three. The process will be to select sections from the raw cloth and place them next to each other, in conformity with an overall design or template. This links the sections and gives overall coherence. The template is the extract quoted above, from the writings of Rosemary Ruether. In a sense, this article is a meditation on the extract, and on the last sentence in particular. The entire piece is framed around a metaphysic of relationality. This is the border. The structure, then, is constructive. It proceeds in a step-by-step

3. See Lorna Hallahan's essay in Chapter 6 in this volume, which details the dynamics behind repair.

4. Grace Jantzen, "Flourishing," in *An A to Z of Feminist Theology*, ed. Lisa Isherwood and Dorothea McEwan (Sheffield: Sheffield Academic Press, 1996) 70–72.

5. Jantzen also argues that the model of salvation traditionally conceived is inherently individualistic and can become introverted and depoliticized. While it is possible to interpret salvation in this way, there are alternatives. See, for example, John Zizioulas's understanding, as described by Patricia Fox in Chapter 5 of this volume. Zizioulas claims that salvation is identified with the realization of personhood. Salvation interpreted by liberation theologies is far from being introverted and depoliticized. I am indebted to Patricia Fox for these insights.

manner, in order to respond to the question: How can I heal so that flourishing can take place?

Quilts, while works of art, also have a practical dimension. They offer warmth and protection. Similarly, this theological quilt must be feasible; it must service everyday needs and include a justice dimension.

First the border must be decided upon. Its dimensions determine the size and the function of the piece. In other words, the fundamental place of relationality must be established.

Relationship is essentially about connection. It is encapsulated in a move from a metaphysic centered on being or substance to one centered on dynamic relationship. The denial of relationality has its roots in attitudes such as individualism, objectification, and traditional notions of power, which result in domination and control. These postures distort true or right relationship, and their enduring influence is indicative of the paucity of Western culture. In particular, feminist theologians cite a blindness to the pervasive and powerful effects of what Catherine Keller terms "the separative sensibility."[6]

The separative sensibility divides human from human, human from nature, and God from both. It repudiates the relationality which is the basic element of existence and which "extends far beyond what our consciousness is capable of absorbing."[7] Under the influence of process philosophy, Marjorie Suchocki can suggest that relationships constitute existence. Entities are not enduring substances but are "events" where the universe composes itself:

> Relational theology attempts to delineate the underlying nature of the world. . . . Each reality receives from all that have preceded it, and gives to all who succeed it . . . . Thus relationships are not secondary characteristics of things; rather in and through relationships each thing becomes itself. Relationships are constitutive of existence, and therefore through relationships all things are woven together.[8]

McFague adds a further critical element: non-human nature is as worthy of subjecthood as humans. As relationships are only possible between subjects, the subject-subject model needs to be extended to include

6. Catherine Keller, *From a Broken Web: Separation, Sexism, and Self* (Boston: Beacon Press, 1986).

7. Mary Grey, *Redeeming the Dream: Feminism, Redemption and Christian Tradition* (London: SPCK, 1989) 30.

8. Marjorie H. Suchocki, *The Fall to Violence: Original Sin in Relational Theology* (New York: Continuum, 1995) 104. Process theology developed from Alfred North Whitehead's "Philosophy of Organism" and Charles Hartshorne's theology. There are various schools of process thought and process theology. For Suchocki, relational theology and process theology are synonymous.

non-human nature.[9] To treat nature as subject as opposed to object is to simultaneously confer value on it and to bring it into kinship with humanity.

According to science, we are interconnected with other entities in the biosphere. Physical, biological, and chemical processes indicate that we are in relationship with the elements, plants, species, and physical forces in undeniable ways. However, relational language used in regard to nature also belongs to another realm—that of imagination and experimentation, images, metaphors, and models, such as McFague's. Relational language can be a powerful ally in informing action. It depends on concepts such as respect, reciprocity, and mutuality and relies on the particular, concrete, and immediate. Values such as empathy, compassion, and care are paramount. Such intuitions pervade the work of the women writers considered here. Relational thinking starts with "our bodies, ourselves."[10] It is where we encounter our capacity for creativity and interaction, for meeting the other subject, human or non-human, and thereby for releasing the possibility for our, and their, transformation.

Relationship, then, is not only the border; it is a theme that recurs again and again. As a structure it underpins all of what follows. Reality (which must include humanity and all earth others) is relational. God is relational.[11] That is where the quilt begins.

* * *

If God is thought of as present in creation by bringing forth
all things in life-giving interrelations
and by a constant process of renewing life-giving relations . . .

9. The subject-object model, which has been in ascendancy since the Enlightenment, is destructively hierarchical, individualistic, utilitarian, and blind to the limitations of nature. It is characterized by a dualism between knower and known, named by McFague as "the arrogant eye." The arrogant eye is acquisitive, seeing all in terms of self. It simplifies in order to control, denies difference and mystery, and has its foundations in patriarchy. Nature is primarily a spectacle. Conversely, "the loving eye" is its precise opposite, where subjecthood is valued and conferred to earth others. The loving eye belongs to the ecological paradigm, which is deeply relational and eminently Christian. Sallie McFague, *Super, Natural Christians: How We Should Love Nature* (Minneapolis: Fortress Press, 1997) 67–117.

10. Beverly Wildung Harrison, "The Power of Anger in the Work of Love: Christian Ethics for Women and Other Strangers," in *Feminist Theology: A Reader,* ed. Ann Loades (London: SPCK, 1990) 203.

11. For an interesting and fruitful contrast to this essay, see Patricia Fox's essay in Chapter 5 of this volume, "God's Shattering Otherness: The Trinity and Earth's Healing," where she carefully details the relevance of the relational theologies of John Zizioulas and Elizabeth Johnson applied to God as Trinity.

Feminist theologians allege that instead of fully respecting the God present in all things, classical Christian theology has been shaped in a crucible where to transcend is to separate. Consequently, the devaluing of immanence is set, and God becomes a removed, separate deity who is "wholly apart" and "over against" (Heyward), the "ruling male" (Grey), the "masculine self writ large" (Keller). As Ivone Gebara succinctly puts it, "I am suspicious of this omnipotent god, this self-sufficient god, this god beyond the earth and the cosmos . . . the celestial 'double' of powerful men."[12] This God sits comfortably alongside the cultural dominance of patriarchy. Feminist theologians, along with others, sense that it is imperative for God to be wrestled to the ground from the above and the beyond, to be much more within, infusing everyday life with the presence of the sacred.

It is not transcendence *per se* that is objected to as much as the separative sensibility at work in many conceptions of transcendence. Heyward remarks that the separative God subjects humanity to an existence characterized by estrangement and isolation.[13] For Rita Brock, it makes no sense that doctrines of salvation make separation and disconnection the source of reconciliation and connection. The challenge, remarks McFague, is to "rethink transcendence in immanental ways."[14]

One move is to examine nature itself for the source of transcendence. McFague, in this vein, declares that all bodies are alive with the breath of God. God's transcendence is embodied in the entire universe, as well as in each particular being. God is the source of all the incredible fecundity and variety of life.[15] "The world as God's body" is a radicalization of divine immanence, for God is not present to us in one place only (Jesus), but in and through all bodies, from subatomic to galactic ones, in the sun, trees, animals, and rivers. All bodies, she claims, should become part of the language we use to speak of God.

In a similar concentration on the diversity of living entities, Suchocki suggests that transcendence lies in the *uniqueness* of each entity in nature.

12. Ivone Gebara, "The Face of Transcendence as a Challenge to the Reading of the Bible in Latin America," in *Searching the Scriptures: A Feminist Introduction,* ed. Elizabeth Schüssler Fiorenza (New York: Crossroad, 1993) 1:175.

13. Carter Heyward, *The Redemption of God: A Theology of Mutual Relation* (Lanham, Md.: University Press of America, 1982) 1.

14. Jürgen Moltmann uses the phrase "immanent transcendence" as the lynchpin of his pneumatology. See *The Spirit of Life: A Universal Affirmation* (London: SCM, 1992). Other ways explored by theologians include developing the concepts of God's Shekinah, Wisdom of God, and God's indwelling Spirit. See also Denis Edwards's essay in Chapter 3 of this volume.

15. Again, McFague is not unique in this. See Denis Edwards's work in *Jesus the Wisdom of God: An Ecological Theology* (New York: Orbis Books, 1995) and *The God of Evolution: A Trinitarian Theology* (New York: Paulist Press, 1999).

For Gebara, the transcendence is in the relations themselves: "We must reaffirm transcendence as relation, relation present to everything, relation articulated among all human beings, animals, plants, earth, air, fire, water, cosmos."[16]

Heyward invokes incarnation, which has the effect of concentrating attention on bodies and matter:

> Relational theology is incarnational. It has to be. Relation is incarnate, between us, in the physicality of all we do: breathe, move, think, feel, reach, touch. The God of whom I speak, the resource and power of relation, that with whom we image ourselves in relation, is the flesh *(sarx)* alive in human being, active in human life, on earth, in history.[17]

Grey argues that transcendence and immanence are mutually enriching dimensions.[18] Immanence is a "rich, many-layered interiority," where the divine energizes our thirst for meaningful relating and our passion for justice. Immanence is the "delightful assurance" that God is present in particular concrete situations, while transcendence means that God takes the initiative in the creating and redeeming activity. God, she concludes, is the healing energy at the heart of the universe.

This healing energy need not be confined to the healing of nature. As Melissa Raphael points out, "if God is immanent in the erotic processes of life, then the maintenance and healing of these processes, which are so often damaged by patriarchy, is also a healing of humanity's relationship with God."[19]

* * *

. . . . then humans are most expressive of God's presence by manifesting this creative and healing process.

My place is as a *natural being,* interconnected with other natural beings physically, emotionally, and socially. I am, additionally, world-related and other-related, to such an extent that I am *constituted by my relations.*

I have multiple relationships with many others in continuity with myself.[20] All entities are connected on a continuum of life. As McFague argues,

---

16. Gebara, "The Face of Transcendence," 178.

17. Heyward, *The Redemption of God,* 32.

18. Mary Grey, *The Wisdom of Fools? Seeking Revelation for Today* (London: SPCK, 1993) 113.

19. Melissa Raphael, *A to Z of Feminist Theology,* 200.

20. In common with many feminist theologians, it is possible to see the influence of the work of Carol Gilligan and Nancy Chodorow, who suggest that women define themselves through relationships of care and responsibility; likewise, the theological work of Valerie

it follows that I am, in reality, more similar to than different from non-human nature. I may be distinct, but I am not a separate being cleanly divided from the surrounding world, essentially remaining the same self from moment to moment. I do not decide to form relationships but am, from prenatal existence, created in relationship. As Keller puts it, "perichoresis begins at home. . . . For good or ill I become part of you, you become part of me."[21] Grey expresses it this way, "Just as I gather the world into me, so I too am gathered into world-becoming, with my whole relational nexus."[22]

As a human being, I am responsive, fluid, and open, influenced by internal and external relations. My interdependence with others is the basis of my physical and emotional life. However, feminists also point out that as a woman, I am particularly susceptible to losing my identity in a sea of a multiplicity of relations. To McFague, the trick is to develop a concept of individuality that avoids the twin dangers of hyperseparation and fusion, that is, to allow intimacy *and* difference, or to differentiate *within* relation.[23] Brock points out that fusion is not intimacy and connection but, in fact, its antithesis, operating as it does under the false logic that two are really one. The distinctive unique self of the other cannot be experienced clearly in and for itself.

Keller's solution to this dilemma lies with women themselves. Borrowing a theme from Luce Irigaray, she says: "If we consciously claim our participation in all things, or the way the Other is part of ourselves, we may indeed be protected from dissolution by connection . . . if I take the many into myself am I not many?"[24] Keller suggests that relationships can be empowered by ensuring that they are connections that count, and in this way women can be faithful to the "complex of inner and outer connectivity" that

---

Saiving and Judith Plaskow, who unmask the differences between male definitions of sin as pride and women's sins as lack of self esteem, loss of identity in relationship, and self-sacrifice. See Carol Gilligan, *In a Different Voice* (Cambridge, Mass.: Harvard University Press, 1982); Nancy Chodorow, *The Reproduction of Mothering* (Berkeley: University of California Press, 1979); Valerie Saiving, "The Human Situation: A Feminine View," in *Womanspirit Rising: A Feminist Reader in Religion,* ed. Judith Plaskow and C. Christ (San Francisco: Harper and Row, 1979); Judith Plaskow, *Sex, Sin and Grace: Women's Experience and the Theologies of Rienhold Niebuhr and Paul Tillich* (Lanham, Md.: University Press of America, 1980).

21. Catherine Keller, "The Theology of Moltmann, Feminism and the Future," in *The Future of Theology,* 147. "Perichoresis" is a word used to describe the mutual, dynamic indwelling of the Trinity, whereby each Person of the Trinity is present to each other in a shared life.

22. Mary Grey, "Claiming Power in Relation: Exploring the Ethics of Connection," *Journal of Feminist Studies in Religion* 7 (Spring 1991) 15.

23. McFague, *Super, Natural Christians,* 104–105.

24. Keller, *From a Broken Web,* 163.

they sense as integral to themselves. A woman is "always several" but is prevented from dispersion because "the other is always within her." Grey's answer is to allow diversity in all forms to emerge and to be celebrated. For McFague, domination and fusion are unnecessary because God, in grace, accepts me as I am.

The ideal, then, is not the elimination or domination of the other, but "respect for and commitment to the well-being of the other."[25] To desire above all their well-being is to participate in God's own care for the world. Recognizing my place must inevitably lead to adopting a certain attitude toward others, human or non-human. Even though I am constituted by my relationships, I must also respect the integrity and otherness (the subjecthood) of the other with whom I am in relationship. Given that I am fundamentally relational, in what way are creativity and healing integral to my essence as a human being?

In relationship, the other must be allowed to emerge. This is not projection but encounter. It is a giving of my full presence to the other. Openness to the other is to be vulnerable, but it also discloses the possibility of being enriched. While wanting to avoid a kind of simplistic romanticism, it is possible to acknowledge that non-human nature can, in a myriad of ways, enable me to flourish.

Healing is in the co-creation of each other in relationship, but as Martin Buber highlighted, it also lies in the Between, that is, the concentration should not be on me (the self) or on the other (nature), but rather on the creative interaction *between the two*.[26] This is the "more than," the "excess," the "creative middle" that is the unknown and unknowable part of the encounter. In the Between is respect for otherness, respect for the encounter, and respect for our interdependence. This Between cannot be prescribed but is rather the *surprise* in regard to our relations with nature. Such relationships are dependent on intimacy. Intimacy in turn relies on mutuality, reciprocity, presence, and immediacy.

The creativity that leads to transformation in relationship comes from an attitude within, but there is also a mysterious element that one cannot determine. As Buber would say, will and grace must be present. The origin of such grace in an encounter can only be guessed at. Possibly it is from God, possibly from the openness of the other in encountering me. Nevertheless, in this regard I am powerless to force an encounter.[27]

25. Grey, *The Wisdom of Fools?* 75.
26. I am heavily relying on the philosophy of Martin Buber. See *I and Thou,* trans. Ronald Gregor Smith (Edinburgh: T&T Clark, 1958).
27. For an interesting contrast to this rendering of Buber's I-Thou relation, see Lorna Hallahan's essay in Chapter 6 of this volume, concerning Miroslav Volf's "drama of embrace."

Intimacy, with human or non-human, is contingent upon knowledge and appreciation—what Sallie McFague terms seeing with a loving eye as opposed to an arrogant eye.[28] In-depth knowledge of a particular life-form or landscape, for example, includes scientific knowledge of ecological processes along with the kind of intuitive understanding that is built up over time. That includes awareness of change, of points of vulnerability, of moments of communicable beauty or horror. It should be said that such encounters with nature are not confined to the wilderness or rural areas. Nature, composed of living entities, has the ability to respond to my openness to relationship wherever I find myself, but this response is not determined by me. Importantly, it includes the possibility of nature not responding, of being indifferent to my presence, indeed of doing me harm.

Nevertheless, my primary directive is to be with nature, with the other. Being with is something to be endured, even when I encounter the refusal of relationality. The women at the foot of the cross manifested this being with, as did Jesus throughout his life. My *being with* inevitably leads to *suffering with,* to the sharing of another's pain. Such creative empathy is so much a work of God that it is appropriately called an immanent transcendence.

* * *

Sin breaks this creative and healing process . . .

Sin is the antithesis of creativity and healing and distorts true relationality. Sin resists working toward the well-being of others. For Marjorie Suchocki, sin is not primarily rebellion against God, but rebellion against the *well-being of creation.* She redefines sin as "participation through intent or act in unnecessary violence that contributes to the ill-being of any aspect of the earth or its inhabitants."[29]

To be human is to be entangled in sin. For Suchocki, sin as violence comes before pride, self-centeredness, unbelief, or anxiety. Violence is the disregard for the well-being of the other. This denial of the other is a denial of relationality. All humans sin, and in a relational world, the effects of sin are felt by everyone.

Brock agrees. Sin is "a symptom of the unavoidable relational nature of human existence through which we come to be damaged and damage others."[30] Hence sin is a sign of our brokenheartedness, of how damaged we are, not of how evil, willfully disobedient, or culpable we are. Sin is not something to be punished but something that requires healing.

28. See note 9 above.
29. Suchocki, *The Fall to Violence,* 12.
30. Rita Nakashima Brock, *Journeys by Heart: A Christology of Erotic Power* (New York: Crossroad, 1988) 7.

Sin is also the refusal to recognize my place. For McFague, if we live disproportionately, falsely, or inappropriately, we are sinning. Sin is a failure to see God in everything, to see the creative energy that pulsates through all existence. As Grey maintains, God's compassionate presence is an energy driving toward just connections and is the basic rhythm of creation. Sin and evil, however, are the reverse energies, causing the uncreating of the self, communities, and natural systems. Thus structural evil is an unmaking, the reverse of the nurturing tenderness that called creation into being. It is intertwined with "the separative sensibility." Finally, sin is not to accept that one is empowered. To Heyward, sin and evil are the results of *fearing our capacity to relate and to create.*[31]

* * *

*. . . and creates distorted relationships of domination
and exploitation . . .*

I may have the capacity to create, but I also have the capacity to destroy. As Beverly Harrison puts it:

> We do not yet have a moral theology that teaches us the awe-ful, awe-some truth that we have the power through acts of love or lovelessness literally to create one another . . . . Because we do not understand love as the power to act-each-other-into-well-being we also do not understand the depth of our power to thwart life and to maim each other.[32]

Even a small element of domination in a relationship, albeit benevolently cast, reduces the presence of the other and thereby diminishes the creativity of connection and denies the possibility of the healing that leads to flourishing.

Power is distorted not only by individuals but also by systems and structures endemic in society. The ecofeminist critique lends a great deal to a picture of the distortion of relationality. For example, ecofeminists have highlighted the almost universal acceptance of the conventional structuring of relationships in terms of power and submission. Patriarchy, hierarchical thinking, and dualisms aggrandize one side of a relationship to the detriment of the other and thus distort power.[33] The ecofeminist project aims, in part, to disclose multidimensional oppressions that are ideologically supported by phenomena such as sexism, racism, cultural imperialism, and androcentrism. The argument runs that patriarchal culture constructs inferior "others" and

---

31. Heyward, *The Redemption of God*, 159–162.
32. Harrison, "The Power of Anger," 203.
33. See, for example, ecofeminist classics such as: Susan Griffin, *Women and Nature: The Roaring Inside Her* (New York: Harper and Row, 1978); Elizabeth Dodson Grey, *Green*

uses this inferiority to justify their oppression. In order to maintain control, those in power exercise a combination of physical force and psychological domination.[34] As Keller puts it, "Assuming the dominative-possessive view of power invariably sets God up in an abusive relationship to the world," and additionally, she might have remarked, humanity's relationship to the world.[35]

We have come to believe that power is demonstrated by dominance, status, authority, and control. It is a unilateral rather than a shared or relational power. As Brock says, this may be the power I know but it is not the power I am born with.[36] I am, in truth, empowered, but I can only exercise a *certain kind* of power. The power I am born with is not the power to dominate, and it is not necessarily even the power to be active in regard to nature. Activity *may* be required (for example, in terms of restoration and reparation), but a letting be or openness to the inherent healing capacities of nature may be equally appropriate. Each particular encounter I have with nature (human and non-human) is unique. I must be open to the possibility of behaving differently, according to what is required to enable flourishing and according to what emerges from the encounter.

Feminist theologians are universally vocal in stressing that my empowerment means, above all, a hunger for justice. In a relational world, it is impossible to separate justice for nature from justice for humans. To Heyward, right relation, love, and justice are identical. Justice is the result of human passion. This manifests itself in a willingness to endure the pain, fear, pleasure, ambiguity, tension, and transitions of relationships. The connection between creativity and healing is justice. A divine work is to heal the wounds of dysrelation, that is, to exercise erotic power.

For Brock, erotic power is the fundamental power of life, "the power of our primal interrelatedness."[37] It is born into us, heals, empowers, and lib-

---

Paradise Lost (Wellesley, Mass.: Roundtable Press, 1979); Val Plumwood, *Feminism and the Mastery of Nature* (London: Routledge, 1993); Carolyn Merchant, *The Death of Nature: Women, Ecology and the Scientific Revolution* (San Francisco: Harper and Row, 1980); Leonie Caldecott and Stephanie Layland, eds., *Reclaim the Earth: Women Speak Out for Life on Earth* (London: Women's Press, 1983); Carol Adams, ed., *Ecofeminism and the Sacred* (New York: Continuum, 1994); Vandana Shiva, *Staying Alive: Women, Ecology and Survival in India* (London: Zed Books, 1988); Karen Warren, ed., *Ecological Feminism* (London: Routledge, 1994); Mary Mellor, *Feminism and Ecology* (Cambridge: Polity Press, 1997); Ann Primavesi, *From Apocalypse to Genesis: Ecology, Feminism and Christianity* (Tunbridge Wells: Burns and Oates, 1991); Rosemary Ruether, *Gaia and God: An Ecofeminist Theology of Earth Healing* (London: SCM, 1993); Rosemary Ruether, ed., *Women Healing Earth: Third World Women on Ecology* (London: SCM, 1996).

34. It is important to note that ecofeminists do not seek to establish these connections themselves but merely to observe what has already been instituted under patriarchy.

35. Catherine Keller, "Power Lines," *Theology Today* 52 (1995) 189.

36. Brock, *Journeys by Heart*, 25.

erates. Its manifold forms create and emerge from the heart, that passionate mystery at our very center. Eros is a transformative, whole-making wisdom that emerges with the subjective engagement of the whole heart in relationships. Erotic power creates and sustains connectedness, intimacy, and generosity. It is the energy that produces creative synthesis. Erotic power integrates all aspects of our life, making us whole. It also resides in the matrices of our connections to our bodies, to others, and to the earth. Through it we experience the richness of our lives and the flourishing of the world.

Grey, Heyward, and Brock agree that such power comes from God. For Heyward, it is imperative that humanity respects its own *power to redeem.* Decisively, this creative power comes *only in relation,* in the intimate and immediate power between self and the other. Thus humanity and God are *co-creative agents* who voluntarily participate in making right relation here and now.[38] As a relational process, empowerment involves mutual change and growth. This power is the root of love and justice; it is not only from God, it *is* God. When we are empowered and empowering creatures, we ourselves are capable of "godding." What Heyward calls "to god in the world" is, in reality, humanity creating right relation. Verbs are appropriate here. They suggest animation. Likewise, Grey focuses on the energy of God: "Power in this context means the relational drive and energy which both empowers human becoming and ecological growth and is the locus of Divine presence in the universe."[39]

Heyward expresses what others also acknowledge: Jesus exemplified power and intimacy in mutual relation; he showed us how to be empowered to "create the world."[40] He demonstrated that to sustain relation is also to suffer the pain and trauma of broken relation. Harrison expresses our vocation in these terms: "Like Jesus we are called to a radical activity of love, to a way of being in the world that deepens relation, embodies and extends community, passes on the gift of life . . . the power to give and receive love . . . is a tender power."[41]

The final piece of the quilt is the central panel. It is the light, in contrast to the dark sections concerning sin.

* * *

37. Ibid., 26.

38. Heyward, *The Redemption of God,* 1.

39. Grey, "Claiming Power in Relation," 14.

40. Heyward makes the distinction between Jesus' power as *dunamis,* raw, spontaneous, charismatic power, rather than *exousia,* the conferred authority characterized by domination and hierarchy.

41. Harrison, "The Power of Anger," 210.

Salvation is the process of conversion from distorted relationships and the regrounding of selves in healing relationality.

Salvation is God's healing work, a sign of grace. God's intimate and integral relationship to me as a living being means that salvation is *also* brought about when I reground myself in the healing relationality. That is, to acknowledge my part in a continuous process of creative transformations toward the well-being of all forms of life is also to enrich the life of God. Healing, ecological or otherwise, comes about through this series of transformations, which are a co-creation of Creator and creature. I am involved in an ongoing process that is dependent on God but is also contingent upon me knowing my place, upon compassion, passion, sometimes active and sometimes passive participation, and mutuality in my relations with nature. I do indeed have responsibility, both the capacity to respond and the ability to effect this creative transformation (to use Suchocki's play on words).[42] This form of empowerment is not anthropocentrism, because this power does not originate from me alone but from the source of all relational power, God.

Forgiveness for my part in the damage can be requested but never assumed. As Suchocki says, God is the ultimate context for forgiveness.[43] Forgiveness can break the cycles of sin and violence that characterize distorted relationships. Even as a violator, I am invited into the great transformation that is God.

Thankfully, the healing that I am empowered to effect is not a solo activity to be done in isolation. Not only is it God's work, it is the also the work of others, with whom I am interrelated and to whom I can turn for mutual support. Heyward calls this the path of "mutual messianism."[44] As she says, messiahs encourage us beyond our hesitation when faced with claiming the power of right relation.

* * *

This quilt establishes a basis for understanding my responsibility for the state of the world. This responsibility is very near me. In a relational world I am not a passive recipient, resigned to accept my own and the world's lack of flourishing, but can, with God's grace, heal.

---

42. Suchocki, *The Fall to Violence*, 132.

43. Suchocki defines forgiveness as the willing of the well-being of both the victim and the violator, in the full knowledge of the circumstances of a violation. It is a matter of the intellect/will rather than the emotions. *The Fall to Violence*, 144.

44. Heyward, *The Redemption of God*, 165.

Barramundi

When the tide goes out
All the men and boys go fishing.
We fish barramundi.
We find 'em in little pools
And in big pools
With stingrays and bluebone and others.
We spear 'em quick
And then the women cook them
All sorts of ways . . .
You wrap 'em in foil
Chuck 'em in the coals
Or make fish soup
Or *numus*—that's raw fish.

You see,
We are a saltwater people,
For generations my people live there.
There's sacred places,
Places we go,
Places we can't.
This is our saltwater dreaming.

So we fish the barra . . .
We fish enough for a feed.
We don't need to get more
Then they're always there.

William Hewitt

Chapter 9

A Timely Reminder

Humanity and Ecology in the Light of Christian Hope

Gregory Brett, C.M.

Introduction

To be human is to live in hope. It is intrinsic to the human condition and is not something peculiar to people calling themselves religious or Christian. But what is the foundation of our hope? Ernst Bloch argues that the possibility of hope arises out of the human capacity to experience the "not yet" of life. This human capacity to anticipate, which often works through our imagination, stimulates hope within the human heart. It is the seed in which what is to come is set, and hence the kernel of real anticipation.[1] It can be said to be the foundation of the human ability to dream, to hunger for the more.

The Christian perspective on hope is grounded in the passion, death, and resurrection of Christ and the sending of the Spirit. The Christ-event is the heart of Christian hope. The area of theology that reflects systematically on hope is called "eschatology."[2] Planted deep in Christian consciousness is the belief that life is irrevocably changed but not ended. Teilhard de Chardin wrote about this belief in terms of the transforming presence of the

---

1. Wolfhart Pannenberg, *Systematic Theology* (Edinburgh: T&T Clark, 1998) 3:175.

2. A significant theological reason why eschatology has taken center stage in current theology has been the retrieval in biblical studies of the Reign of God as the central message of Jesus. The faith of Jesus and his followers was steeped in the expectation of the coming of the Reign of God. Their reality was saturated with hope and promise. The authentic life of faith was one of looking to the fulfillment of God's promise, based on complete trust that God is a promise keeper. And it was a promise that involved all God's creation.

cosmic Christ to creation and humanity.[3] The life-giving, transforming spirit that was released in and through the Christ-event cannot be stopped. Paul describes the whole of creation as groaning in one great act of giving birth (Rom 8:18-28).

The challenge before us is to allow that hope-filled vision to say something to the major questions confronting us today. In this short essay I will show that this vision of reality (eschatology) is relevant to the present ecological crisis. I will suggest that it has a central place in guiding our thinking and behavior as we grapple with a world that has been severely damaged by the actions and self-interest of humanity. Eschatology has an important contribution to make in ecological theology.

Ecological theologies today range from "new cosmologies"[4] on the one side of the spectrum to Evangelicals for Social Action on the other.[5] In between these positions there are a number of others that build upon rich scriptural resources and develop a systematic theology for a responsible ecological attitude. These include Walter Brueggemann on the land, Jürgen Moltmann on the Sabbath, Elizabeth Johnson on the indwelling Spirit of Wisdom, Denis Edwards on Jesus as the Wisdom of God, and Douglas Hall on stewardship.[6] These theologies hope to inspire us to ac-

---

3. Teilhard de Chardin, *The Divine Milieu* (New York: Harper and Row, 1960).

4. The term "new cosmologies" attempts to capture the central idea in a number of Protestant theologies that promote human identification with nature. These theologies appear to meld with New Age doctrines and in some places go as far as advocating worship of the earth. For further discussion on these theologies, see Ed Hindson, *End Times, the Middle East and the New World Order* (Wheaton, Ill.: Victor Books, 1991).

5. These Protestant theologies focus on New Testament evidence (as distinct from Old Testament reflection) and argue that Jesus' response to problems was done within a political framework that got results. They argue that far from being a theorist, Jesus was a practical person who undertook change. They suggest that this is the model for today's Christian environmentalists. See Vernon Visick, "Creation Care and Keeping in the Life of Jesus," in *The Environment and the Christian: What Can We Learn from the New Testament?*, ed. Calvin B. DeWitt (Grand Rapids, Mich.: Baker, 1991). See also Wesley Granberg-Michaelson, *Worldly Spirituality: The Call to Redeem Life on Earth* (San Francisco: Harper & Row, 1984).

6. Walter Brueggemann, *The Land: Place as Gift, Promise and Challenge in Biblical Faith* (Philadelphia: Fortress Press, 1977); Jürgen Moltmann, *God in Creation: A New Theology of Creation and the Spirit of God*, trans. Margaret Kohl (San Francisco: Harper and Row, 1985); Elizabeth Johnson, *Women, Earth and Creator Spirit* (New York: Paulist Press, 1993); Denis Edwards, *Jesus the Wisdom of God: An Ecological Theology* (Homebush, N.S.W.: St. Pauls, 1995); and Douglas Hall, *Imaging God: Dominion as Stewardship* (Grand Rapids: Wm. B. Eerdmans, 1986). Over the last thirty years there has been a "greening" of theology. For developments in Protestant theology, see Robert Booth Fowler, *The Greening of Protestant Thought* (Chapel Hill, N.C.: University of North Carolina Press, 1995) 39–44. For developments in Catholic theology, see Sean McDonagh, *The Greening of the Church* (Maryknoll, N.Y.: Orbis Books, 1990).

tion by awakening our sense of kinship with the earth and with all living things. They hope to deepen our gratitude for the realm of nature and our love for other creatures by evoking in us a heartfelt reverence for the whole of God's creation.[7]

An eschatological theology in dialogue with the questions of ecology today aspires to this same hope by returning to a fundamental theme in the Scriptures, that of the divine promise for future fulfillment, and asks: Is it of any relevance here? Can the hope-filled dimension of Christianity, founded on the ancient Hebrew experience of God's promise and fidelity and grounded in the incarnation, become the keystone of an environmentally sensitive religious vision? Most recent attempts by Christians to build an environmental theology have made only passing reference to the eschatological dimension of ecotheology.[8]

I would like to make a brief comment about language. I agree with Moltmann that the term "nature" should be removed from theological discourse. He encourages us to use the term "creation" rather than "nature." To approach the world as nature suggests that it is radically different from human beings and, by extension, has little or nothing to do with human salvation. To approach the world as creation may suggest a more inclusive appreciation of the human and non-human dimensions of reality. So where possible I will use the term "creation" as an inclusive theological term.

Eschatology and Ecology

There are those who are critical of an eschatological approach. It appears to some of its critics like stories from the beyond and "news from nowhere." I offer two examples, one from outside Christianity and one from within it. Australian philosopher John Passmore argues that Christianity is limited in what it can say and do in the current ecological challenge. He believes that there is an ever-present dualism in Christian thought that drives a wedge between God and nature. Passmore doubts that Christian theology can ever reshape itself in an ecologically helpful way without ceasing thereby to be Christian. He argues that Christianity's traditional longing for another world is too ingrained and pervasive in its theology. Further, he says, this Christian emphasis on another world (which he associates with eschatology) actually sanctions hostility toward nature. He believes that we must come to the realization that we are all

7. Harold Wells, "Christological Implications for an Ecological Vision of the World," *Toronto Journal of Theology* 15 (Spring 1999) 56.

8. Major exceptions are Jürgen Moltmann, *God in Creation,* and John Haught, *The Promise of Nature: Ecology and Cosmic Purpose* (New York: Paulist Press, 1993). It is present but not explicit in Denis Edwards, *Jesus the Wisdom of God.*

alone in this universe. There is no one and nothing to assist us except ourselves. The world is in fact wholly indifferent to our survival. He therefore
proposes that the only healthy alternative to the Christian approach is a
radical secularism.[9]

American theologian (or self-styled geologian) Thomas Berry argues
that the recurring emphasis on the future found in the Scriptures has
wreaked ecological havoc. For Berry and many others, the future orientation of the Bible has bequeathed to us the dream of progress, which has
caused us to bleed off the earth's resources while we uncritically pursue
an elusive future state of perfection. Hoping in a future promise can lead
us to sacrifice the present world for the sake of some far-off fulfillment.[10]

The critiques of Passmore and Berry are significant as they challenge
both our understanding of what we mean by eschatology and its validity as
a partner in the development of an inclusive ecological theology. Passmore's
philosophical critique of Christianity is important because he represents a
strong secularist viewpoint that Christianity is not in a position to make a
positive contribution to the ecological challenge. He is convinced that Christianity is too anthropocentric and suggests that this is most evident in its
eschatological teaching. Berry's theological critique of Christianity is important because he suggests that the prophetic dimension of an unfulfilled
future found in the eschatological teachings of the Church shares some of
the blame for the Church's involvement in the present ecological crisis.
Moreover, Berry suggests, only by jettisoning this eschatological dimension
can the Church fruitfully reenter the ecological conversation.

Passmore and Berry offer a harsh indictment of Christianity and
specifically of eschatology. As Christians, we stand accused, at least of ambiguity in relation to environmental questions and in our response to the
ecological crisis.[11] If we were to scan the writings and teachings of the
Christian tradition, we would find that the welfare of the natural world has
received uneven attention. Passmore is partially correct in his observation
that in recent memory Christianity looked vertically above to a completely
different world as the place of fulfillment. The creation and the universe itself had no future. Looking at creation and the universe as the embodiment
of the promise had fallen out of Christian discourse. Only the immortal
human soul could look forward to an ongoing life, value, and salvation.
This salvation was to take place in some completely different domain,

---

9. John Passmore, *Man's Responsibility for Nature* (New York: Scribner, 1974) 39–40,
184.

10. Thomas Berry, *The Dream of the Earth* (San Francisco: Sierra Club, 1988) 204.

11. For a helpful ecological reading of the tradition, see Paul Santmire, *The Travail of
Nature: The Ambiguous Promise of Christian Theology* (Philadelphia: Fortress Press, 1985).

where all connection with "nature" and bodiliness would be erased. This dualistic incision between what was understood as material and what was understood as spiritual excised not only the bodiliness of humanity but also the larger context of our bodiliness, the entire physical universe that is inseparable from our being.[12]

Berry is correct in suggesting that Christianity participated in the destruction of the earth's resources. While Christianity has been a proponent of development and progress for individuals and nations, it has failed to critique this thinking for lacking ecological awareness. In fact, Christianity has often forgotten Paul's intuition that the entire universe yearns for and participates in redemption. A theology of development or progress had submerged the more profound theology of creation. The results of that theology and the havoc it wrought in the creation are clear. To that extent Christian theology stands indicted. However, to lay the blame on the Christian understanding of hope and to further suggest that the theology of hope, or eschatology, has little to offer in responding to the present crisis is to continue to work from a narrow or diseased understanding of eschatology. Further, to sideline it from the theological discourse may prove to be unhelpful.

During the past century we have rediscovered the central place of eschatology in Christian faith. Jürgen Moltmann has consistently argued that all Christian theology must be eschatology. It cannot really be only one part of Christian doctrine; instead, the eschatological perspective is characteristic of all Christian proclamation.[13] Karl Rahner points out that we can only understand what Christian faith is when we recognize clearly the edifice of hope that is built upon it, in which faith accepts God's revelation as promise.[14] Rahner remains critical of all attempts to treat eschatology as if it were simply the caboose at the back of the train of theology, as something simply at the end, having no organic relationship to what comes before and whose detachment from the train would not essentially alter it.[15] Rather, Rahner insists, eschatology must be organically related to all branches of theology. John Haught argues that "to suppress the theme of hope and promise whenever we do any kind of theologizing, no matter what the occasion or the issue, is to fail to engage the heart and soul of this

---

12. The insight that we are made from stardust is just one piece of this larger picture. See Denis Edwards, *Jesus and the Cosmos* (New York: Paulist Press, 1991).

13. Jürgen Moltmann, *Theology of Hope*, trans. James W. Leitch (New York: Harper and Row, 1967) 16.

14. Karl Rahner, "The Question of the Future," in *Theological Investigations* 12, trans. David Bourke (New York: Seabury Press, 1974) 183.

15. Karl Rahner, "Eschatology," in *Sacramentum Mundi* (New York: Herder and Herder, 1968) 2:244.

tradition."[16] Theology must be saturated with hope for the future. If that is so, then ecological theology will need to be grounded in hope for the future. What is critical is how we understand this orientation to the future, this fulfillment of the promise-laden character of Christian faith.

Drawing upon the work of Rahner, I will present three guiding principles that are foundational to an inclusive and balanced understanding of eschatology. I will draw a number of implications from these principles that are integral to the ongoing conversation with ecology.

Three Guiding Principles in Eschatology

Karl Rahner has outlined the parameters of a relevant eschatology, and his insights have something central to say in relation to creation and ecology. At the heart of Rahner's eschatological theology is his profound insight into the Gospel of John and its implications for human life. Rahner understands that as a consequence of the Word of God becoming flesh, the mystery both intimate and incomprehensible has already become the innermost life not only of humanity but of all creation.[17] Rahner's eschatology is understood as an extension of his Christology and anthropology. The Word has become flesh. On this insight he develops three rules or principles that assist us in understanding and applying eschatological statements. His three rules of eschatology have direct bearing on the criticisms proffered by the likes of Thomas Berry and John Passmore.[18]

1. *Eschatology is not simply futurology.* Eschatology has an intrinsic relationship to the present because the end is not added to history like a second story is added to an already completed house, having an essentially superfluous relationship to it. Eschatology is a particular way of speaking of the future from the basis of present experience reflected upon in a religious way. Rahner asserts that humans can only understand themselves and their present in their relationship to the absolute future.

Humanity and creation exist in history. Part of living is to move through time. We (humanity and creation) cannot be understood or find our proper orientation without a view of the past and a view of our future. That involves a looking back and remembering and a looking forward and anticipating in hope. To say that we are a people of memory means that no part of our living or our actions is taken for granted. To say that we are a

16. Haught, *The Promise of Nature,* 105.

17. Marie Murphy, "A Critical Analysis of Karl Rahner's Eschatology," Ph.D. diss. (New York: Fordham University, 1985) 194–195.

18. Karl Rahner, "The Hermeneutics of Eschatological Assertions," in *Theological Investigations* 4, trans. Kevin Smyth (Baltimore: Helicon Press, 1967) 323–346.

people of anticipation or hope means that we understand our present only to the degree that we grasp it as moving toward the future.

Since eschatology is in some way talk about a future (using the language of fulfillment or completion), its inner dynamic may be perceived already at those levels in which we speak of our future in an everyday sense of the word. In everyday language we speak of the future out of our experience of the present. Insofar as the present experience includes the polarities of good and evil, when we dream of a better future we envision overcoming the evil factors in the present and maximizing the good in the present experience. We dream of a better tomorrow by engaging in the realities of today. The development of Christian eschatological language follows in similar vein. The principal difference rests in the fact that the present experience from which it proceeds is the religious experience, focused on God's action in Christ. Thus, as we come to reflect upon our experience, we understand that God, who is the absolute future, is already present in the center of human existence.[19] What happens now and what we do now matters.

2. *Eschatological statements are not to be seen as predictions.* As stated above, eschatology deals with God's own self-communication, which remains essentially mysterious. Hence eschatology cannot have the status of a cosmic weather forecast. Eschatological statements do not give us information about the future before the fact. Eschatology is not a fund of arcane information about the events that are to take place at the end of history. Rahner sees such a detailed reporting of future events as apocalyptic.[20] Although the final salvific event has occurred through Christ and is present now in human reality, human history at the same time is moving into an absolute future that is not capable of being manipulated or planned by humanity and comes as absolute gift. It calls for openness to the approach of the absolute future.[21] Therefore, the purpose of the highly symbolic language of eschatology is to hold us open to an awareness of a future that remains always obscure. It does this so that in the light of the hoped-for future, we may accept our past and present in and with the

19. Karl Rahner, "The Theological Dimension About the Question of Man," in *Theological Investigations* 17, trans. Margaret Kohl (London: Darton, Longman and Todd, 1981) 55f.

20. Rahner, "The Hermeneutics of Eschatological Assertions," 337. Rahner's use of the word "apocalyptic" could well be challenged here in the light of current biblical scholarship. The point Rahner is making is that "to extrapolate from the present into the future is eschatology, to interpret from future into the present is apocalyptic." See also Peter Phan, *Eternity in Time: A Study of Karl Rahner's Eschatology* (Selingrove, Pa.: Susquehanna University Press, 1988) 71.

21. Karl Rahner, "Marxist Utopia and The Christian Future of Man," in *Theological Investigations* 6, trans. Karl H. and Boniface Kruger (New York: Seabury Press, 1974) 60.

world, as a real factor in the actualization of that possibility that was initiated by God in creation.

And because our future in the world and the universe is God, Rahner points out that the "eschatological future remains uncontrollable and hidden and yet also present, something we really look forward to, something in the presence of which we hope, love, trust, and surrender ourselves."[22] We deal with the realities of the present yet remain open to a surprising, mysterious, and overwhelming future.

3. *Eschatology is grounded in the Christ event.* When we speak from an eschatological perspective, we are not speaking about a future that has yet to arrive but a future whose culminating point has been manifested in a historical event, the incarnation. Christ is the norm and foundation of eschatology. Specifically, as Rahner has suggested, it is the present experience of God's grace in Christ.[23] The resurrection of Jesus assures us that life is the principal and overwhelming ingredient in creation and the future of creation. This event reveals and inaugurates the future of the universe for its culmination of God's self-communication—the point at which God irrevocably commits Godself to the world—and this self-communication is the proper consummation of the universe. As Rahner argues, eschatology as a Christian mode of discourse is a projection of fulfillment from the present experience of the human situation insofar as the present is already conditioned by the mystery of the Christ-event.[24]

Implications

In the light of these guiding principles, Rahner makes a number of points pertinent to our discussion on ecology. It would be wrong to think of the absolute future as simply involving the abolition of the present world. The world is not a training ground for souls in the making, which, having served its purpose, can be dispensed with. It is precisely this world that is consummated.[25] It is this world and its history, which constitute the

---

22. William M. Thompson, "The Hope for Humanity: Rahner's Eschatology," in *A World of Grace: An Introduction to the Themes and Foundations of Karl Rahner's Theology,* ed. Leo J. O'Donovan (New York: Crossroad, 1989) 158.

23. Rahner, "The Hermeneutics of Eschatological Assertions," 332–333.

24. Ibid., 329, 335, 342.

25. "The physical world is not merely the outward stage upon which the history of the spirit, to which matter is basically alien, is played out, such that it tends as its outcome to quit this stage as swiftly as possible in order really to achieve full and complete spirituality in a world beyond that of matter." Karl Rahner, "Immanent and Transcendent Consummation of the World," in *Theological Investigations* 10, trans. David Bourke (New York: Seabury Press, 1977) 285.

connatural setting of human fulfillment, which is "the dimension in which the final consummation and the absolute future are made real."[26]

In spite of what Passmore, Berry, and others may assert, eschatology does not look toward a spiritual heaven unconnected to the earth and its history. Eschatological promise is not tied to an other-worldly inheritance for human souls that bears no resemblance to the present state of affairs of this world and the creation. Rather, we are speaking about promise as the culmination of the entire story of the universe, which includes human history but is not reduced to human history. To view the divine promise in this way broadens our horizon of what is included in this promise. Nothing is lost in the absolute future of God. The beauty and complexity of the world that is and the beauty and complexity of the world that has passed are never completely lost but are taken up into God's own life. The divine presence and glory of God (the *shekinah* of God) includes the everlasting preservation of the transient beauty that makes up every phase of the long story of the universe. If the idea of God means anything to people of faith, it means that loss, perishing, and death are not the last word about reality.[27]

However, Christian hope does not generate a new version of Quietism. As outlined in the first guiding principle, the eschatological future toward which the universe moves is not some far-off dream waiting to happen in a fashion completely unconnected to the present. It is a blossoming of the present that will include the present, and all present moments, within itself. Thus our present environmental care is of a piece with that which we wait for in joyful hope. For Rahner, this world is not simply the world that God created; it is also a world shaped by what human beings do. Human beings, in the language of the Second Vatican Council, "project further the work of creation."[28] This means that what the human being does not only has significance in the present moment but will also pass into the consummation of the world.

Hence what human beings do in the world and with the world has eschatological significance. Eschatology does not dissolve human responsibility but rather radicalizes it. Human beings have the capacity and the responsibility of enabling the world to be more open to its absolute future. And that is the precise way in which humans are more open to their absolute future. For humanity and the world, it is an incarnated future.

---

26. Rahner, "The Theological Problem Entailed in the Idea of the New Earth," ibid., 268.

27. Haught, *The Promise of Nature,* 125.

28. "The Church in the Modern World" *(Gaudium et spes),* art. 57. Eschatology does not lead to inactivity. See Karl Rahner, "The Church's Commission to Bring Salvation and Humanisation to the World," in *Theological Investigations* 14, trans. David Bourke (New York: Seabury Press, 1976).

The separation of earth from heaven is ecologically inconceivable. The only wholesome way we can think of heaven today is in terms that include the salvaging, not the discarding, of the earth and its history. The conviction that the world attains an objective immortality by being received into God's everlasting experience has the advantage of underlining the continuity between the present world and humanity's final destiny. The old dualisms are not acceptable.

I have been arguing that the theology of hope, or eschatology, has something relevant to say to us in relation to the present ecological crisis. I have outlined a number of foundational points developed by Rahner for what may be understood as an inclusive eschatology. First, eschatology deals with our present experience as foundational to our future fulfillment. God, who is the absolute future, is already present in the center of human existence. Second, eschatology deals with God's self-communication, which remains essentially mysterious. We remain open to an unexpected and surprising future. Third, eschatology deals with a future whose culminating point has been revealed or communicated in the incarnation of Christ. In principle, all eschatological statements should be translatable into Christological ones. Christ ultimately reveals the destiny of creation. From those foundations I have sketched a number of implications that directly engage eschatology in the present discourse with ecology. I believe that these eschatological guidelines have implications for the ongoing development of an ecological theology.

Three Timely Reminders

Like all theology, ecotheology seeks to persuade. Rosemary Radford Ruether suggests that when we turn our minds to the earth, deeply ingrained patterns of thought and behavior are challenged.[29] Theologians addressing this issue enter the conversation at different doctrinal points: God, incarnation, creation, pneumatology, or in the case of this essay, eschatology. And they address the issue from a variety of theological perspectives. While there are distinct tensions among these theologians, all are seeking to make sense of the Christian understandings of the relationship between creation, humanity, and God. They do so in what John Cobb calls holistic habits of thought that will empower the struggle for integrity and justice of the creation. For the purposes of this essay, I will draw upon three insights (feminist, process, and sacramental) that offer timely reminders to ecotheology, and suggest a further step or timely reminder drawn from Rahner's eschatological principles.

---

29. Rosemary Radford Ruether, *Sexism and God-Talk: Toward a Feminist Theology* (Boston: Beacon Press, 1983). See especially chapter 3.

1. Earth-Grounding

Feminist theology has long been concerned with nature and critical of not only androcentrism but also anthropocentrism. Rosemary Radford Ruether points out that there will be no liberation for women and no solution to the ecological crisis in a world where the primary relationships are driven by domination. The master-servant relationship originating in humanity has had such a detrimental impact on human development and humanity's relationship to the world.[30] As a response to this destructive imbalance, particularly in an ecological context, feminist theology directs us to be critical of various forms of dualism that understand human life to be antithetical to natural life.

Dorethee Soelle reminds us that a theology that seeks to be ecological[31] will not neglect the "dust factor."[32] She argues that it is impossible to divide the world into the spiritual and the "merely material," because matter has a spiritual element and finite spirituality has a material element. The true character of the human being's relationship to nature can be understood only by an "embodied theology."[33] There are three significant features of an embodied theology: (1) the idea that human beings were made from the dust of the ground; (2) the idea that the earth does not belong to humans but that humans belong to the earth; and (3) the conviction that the earth is God's and not a human possession or the possession of some humans.[34] The "dust factor" suggests the end of such dichotomies as body and spirit and human and non-human. It also means the end of an insolent anthropocentrism and wanton domination of creation. However, she warns, human beings are not to regard themselves as passive subjects within creation. That view leads to a devaluation of human effort. Soelle reminds us that human beings are to be

30. Rosemary Radford Ruether, *New Woman, New Earth* (New York: Seabury Press, 1975) 204.

31. See Irene Diamond and Gloria Orenstein, eds., *Reweaving the World: The Emergence of Ecofeminism* (San Francisco: Sierra Club Books, 1990). Common to all ecofeminist theologians is the argument that the connections between the oppression of women and the rest of creation must be recognized and analyzed if either is to be understood properly. Some ecofeminist theory develops this connection between the destruction of the biological community and the oppression of women to include all structures of domination that engender destructive relations. See Rosemary Radford Ruether, *Gaia and God: An Ecofeminist Theology of Earth Healing* (San Francisco: HarperCollins, 1992).

32. Dorethee Soelle with Shirley Cloyes, *To Work and to Love: A Theology of Creation* (Philadelphia: Fortress Press, 1984) 73.

33. For another position on embodied theology more broadly, see Mary Rosera Joyce, "A Revolution in Human Ecology," in *Embracing Earth: Approaches to Ecology,* ed. Albert LeChance and John E. Carroll (Maryknoll, N.Y.: Orbis Books, 1994) 49.

34. Soelle and Cloyes, *To Work and to Love,* 28.

co-creators with God, to participate in a creation that is not yet an accomplished fact. The doctrine of creation is not to be interpreted to mean that God's creative activity ceased at Genesis 3:25. Humanity fulfills itself by fulfilling creation. It does so by participating in God's creative activity. That means that God's power is shared power.[35] This further suggests that the creation is not the raw material for economic development but an essential part of what it means to be human and to fulfill the human vocation.

Soelle's "dust factor" is a timely reminder. It is also an essential part of an eschatological appreciation of reality. Drawing on Rahner's guiding principles, we saw that it is this world and its history that constitute the integral and constitutive setting of human fulfillment. Rahner goes on to say that participating in God's creative activity also means participating in a future that does remain mysterious. Yet our participation is a real factor in promoting the full potential that is present in creation and in us. The full script of what will occur has not been written.

However, our earthling status is well attested to in theology and science. Science has demonstrated that our roots still extend deep down into the earth and fifteen billion years back to the big bang. Hence our hope carries with it the creation's (the whole universe's) yearning for its future. "Billions of years before our own appearance in evolution the universe was already seeded with promise. Our own religious longing for future fulfillment, therefore, is not a violation but a blossoming of that promise."[36] Human hoping is not simply our own construct of imaginary ideals projected onto an indifferent creation, as much modern and postmodern thought maintains; rather, it is the faithful carrying on of the creation's perennial orientation toward an unknown future.

It is a timely reminder that we should not hinder the creation's perennial orientation toward an unknown future. Current thinking suggests that the creation does not grow in a linear or predictable fashion. Most of the things that happen, "from thunderstorms to amoebas, dinosaurs and humans originate in a more spontaneous, 'chaotic,' non-linear fashion out of self-organizing processes insusceptible to exhaustively predictive mathematical assaying."[37] Unpredictability and surprise are hallmarks in the development of life and the universe.

What that means in terms of grounding our ecological concern is that we are obliged to provide every possible opportunity for nature to move toward yet other surprising outcomes, whatever they might be. By damaging and destroying the ecological richness of creation, which has taken

35. Ibid., 39–40.
36. Haught, *The Promise of Nature*, 109.
37. Ibid., 124.

millions of years to develop, we prohibit fresh creation, and in so doing we violate the mysterious future that faith knows by the name of God. In that sense, then, we may think of ecological ethics as flowing quite consistently out of the basic eschatological posture of Christian faith, its openness to surprise. We save the creation because of the incalculable promise it holds. And the future and promise it holds involves us.

2. Earth-Shaping

Process theology has challenged anthropocentric and atomistic views of the world because it is essentially biocentric and stresses the relatedness of all things. This theology calls for a different understanding of the human being's relation to the world from the one engendered by seventeenth-century science. Process thought argues that seventeenth-century science was mistaken in thinking that matter is dead and therefore that nature is lifeless.[38] John Cobb suggests that in Western thinking, creation was seen as a stage upon which the human drama is played out. The stage does not participate in the drama.[39] The creation has no intrinsic value. Process thought denies this and advocates a shift to an ecological attitude that Cobb describes as "respect or even reverence for, and perhaps a feeling of kinship with, the other creatures" in the world of nature.[40]

For process thought, life is best understood not solely in material terms but as occasions of experience of greater or lesser degrees of complexity or richness. All matter, even inorganic matter, is not inert or solid at all but is constituted by "energy events" organized in increasingly complex ways from atoms to humans. "These energy events are capable of having experiences and aims even if only unconscious ones."[41] This suggests that all life has some degree of subjectivity, human subjectivity being the most complex kind that we know. An important consequence of this view is that all life has intrinsic value, even if in varying degrees.

All entities in creation are comprised of intricate relationships with one another. We are reminded today to move beyond substantialist thinking and

38. In the thinking of Copernicus, Galileo, Bacon, and especially Newton, nature was seen as a machine, and machines are not alive and do not possess souls. Bacon went further and spoke of science as the instrument that would force nature to release her secrets. The overriding image was one of torture. Science was to be an invasive instrument in order to attain these secrets.

39. John Cobb, *Is It Too Late? A Theology for Ecology* (Beverly Hills, Calif.: Bruce Publishing, 1972) 81.

40. John Cobb and David Ray Griffin, *Process Theology: An Expository Introduction* (Philadelphia: Westminster Press, 1976) 76.

41. Michael Petty, *A Faith That Loves the Earth* (Lanham, Md.: University Press of America, 1996) 29.

view everything as connected in complex ways to the totality of the universe. The shift in thinking from an understanding of relationship as an accidental connection of substances to the notion that relationship is the very substance of things has been a catalyst in a renewed appreciation of God being all in all. Every entity is in some sense a synthesis of all the relationships presented to it by its environment. There are no substances existing independently of relationships. It is true of God and it is true of God's creation.

If God's creation is one in which all events and phenomena bear some relation to others, humans may no longer define themselves in a simple substantialist manner. The entire universe, in all its depth, duration, and complexity, enters into the production of each one of us. There is no completely impermeable boundary between the human self and its constitutive world.

The observation of process theology that all reality is intricately related is a timely reminder to us. And it has implications for us as we think in terms of eschatology. From an eschatological perspective, humans cannot think of their own destiny, of what they hope for, or of the promise that leads them without involving and being involved with the universe that somehow shares in this same destiny, hope, and promise. Humanity's promise of imperishability and redemption, grounded in the incarnation, is intimately connected with the universe that enters into humanity's self-definition. Thus, in John Haught's words, "If our own personal death does not mean the absolute extinction of our own being, then neither would the transience of the earth or the universe mean its passage into absolute nonbeing."[42]

Eschatology would offer a timely reminder at this point. There is agreement that reality is profoundly relational and shares in the same destiny, hope, and promise as humanity. But how those relationships are maintained, or even more, what they might look like, cannot be predicted. What we do know is they will not be the same. Eschatology, reflecting out of the resurrection experience of Christ, goes beyond the action of God in creation and explicitly affirms the introduction of something qualitatively different, new, and transformed in the fulfillment of the divine promise, the gift of eternal life. Our hope in the future is not simply trust in optimistic development or progress or evolution of the present in an unending time. The process of death and resurrection for humanity and creation is not, as Rahner observes, a matter of changing horses and riding on.[43] The logic of eschatology is not of inference but of imagination.

---

42. Haught, *The Promise of Nature*, 128.
43. Karl Rahner, "Ideas for a Theology of Death," in *Theological Investigations* 13, trans. David Bourke (New York: Crossroad, 1983) 174.

3. Earth-Revealing

Sacramental theology reflects a particular view of God and creation.[44] In terms of ecology, it reminds us that the earth, plants, animals, and human beings are unified by the incarnate Word. The incarnation is not seen as an influence from the "outside" of life but as the ground and goal of the entire world process.[45] And it is because the incarnation is the ground of the life-process that creation has final significance.

Sacramental theology reminds us that the creation and God are not to be confused. The incarnation reveals that the Creator and the creation are not antithetical; rather, creation is a product of grace and is thus a graced reality. It is therefore possible to speak of the creation in the words of Teilhard de Chardin as "holy matter."[46] Nothing in the created world is profane. And it has the potential to reveal the presence of God.

This theology offers a timely reminder that God's relation to creation must be approached carefully so that the mystery of God is respected. As the incarnation is a mystery, in the sense that it cannot be fully comprehended, so God's relationship to the creation is a mystery. However, as in the incarnation the Logos has a real relationship to Jesus' human nature, so God has a real relationship with the creation. This relationship is such that the creation is an expression of God's mystery.

The creation as an expression of God's mystery holds together two theological values. First, creation is seen as God's partial expression. While the creation is certainly not God, God through the Logos is intimately related to the creation and is the dynamic power of all creation's processes.[47] This kind of sacramental theology would be hesitant to suggest that the world is God's embodiment. To do so would be to diminish the significance of the incarnation. And second, God is not alienated from the creation, as that would make the incarnation impossible. Thus creation must be seen as that which God holds in life and is a mystery of "existence." The creation reveals God's "serious play, love of profusion and God's desire to hurl the divine Is! To the outermost galaxies."[48]

Eschatology reminds us that creation is not seen in static terms but as moving toward a fulfillment grounded in the incarnation. In the incarnation the human and the divine, as well as the physical and the spiritual, are

---

44. See especially Albert C. Fritsch, *Renew the Face of the Earth* (Chicago: Loyola University Press, 1987).

45. Ibid., 29–30.

46. Teilhard de Chardin, *The Divine Milieu*, 106.

47. John Carmody, *Ecology and Religion: Toward a New Christian Theology of Nature* (New York: Paulist Press, 1983) 125.

48. Ibid., 120.

seen as joined in a mystical union. That means that the destiny of creation involves a point at which those two converge, though the distinctions proper to each are not annulled. We are further reminded that because creation has not yet arrived at the fullness or fruition of God's reign, it does not merit our worship. It does deserve our profound respect, but not our prostration. If we naïvely invest too much significance in the creation, we will be disappointed by it. It can reveal the divine but not replace the divine. An eschatological appreciation of the creation reminds us that God is the absolute future and prevents an ambiguous approach to creation.

Conclusion

In this essay I have argued that Christian hope (eschatology) has a place at the table of ecotheologies. I have suggested that true Christian eschatology is not stargazing or focused on another, future world. Christian eschatology would not recognize itself in the criticisms represented by Passmore or Berry. Rather, a balanced eschatology has its sights firmly fixed in the present, yet realizes that we are in the middle of unfinished business. The creation is not the dross of matter left behind in the evolution of humanity into some other form. The creation is central to humanity's becoming, as humanity is intimately part of the creation's potential to move into surprising and new expressions.

Karl Rahner offers us three guiding principles from which we might engage questions from an eschatological perspective. His emphasis on Christian hope dealing with the present, the unpredictability of the future because God is the absolute future, yet the faith assurance that life continues grounded in the creation event and the incarnation has given us the theological tools, so to speak, to engage fruitfully in the ecological question.

The dynamic tension of "the now but not yet" that is always present as part of hope, and particularly Christian hope, allowed us, in conversation with feminist, process, and sacramental theologies, to offer some timely reminders that may further the collective goal and hope offered in the theologies of ecology.

It is ultimately hope in the promise of a future fulfillment for the entire creation that can lead us to take suitable actions in the present to save the environment. If we are sincere in proposing a theology of ecology that has connections with biblical faith, we need to make the theology of hope or eschatology a central conversation partner and not a subordinate one in our reflections.

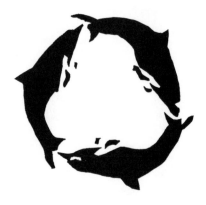

A Delicate Balance

This picture of sea with dolphins and land with snakes speaks about the earth and the sea. They are different, but they overlap each other. They need each other, and share a delicate balance, and if either side is damaged, they both suffer. We can learn from balance—we need others who are different. We should learn from children who are curious and trusting. They don't reject because of difference.

Dolphins are curious, intelligent, childlike, and fearless of outsiders: they reach out for what is different rather than take off when they see it. They move through the water with beauty and grace.

Snakes also move swiftly. They look after their young, but they are feared by outsiders.

Karen Morcom

Chapter 10

Ecotheology as a Plea for Place

Phillip W. Tolliday

Introduction

Some time ago, together with my wife Bev, I attended a fortieth birthday celebration for a friend. In the course of his speech he mentioned that during his life he had moved eighteen times. While driving home after the celebration, Bev marveled that someone could move so frequently. So as a game I began to count up the number of times I had moved and was rather sobered to find that I had lived in no less than nineteen different places. Although Bev's tally was less than mine, she managed double figures without any difficulty at all. It seemed that our friend's experience was rather less rare than we had imagined.

Moving from place to place, we had, each time, experienced a sense of displacement and dislocation. And while common wisdom has it that practice makes perfect, my experience was that each displacement was more difficult than the one that had preceded it. I found all this rather worrying. Was it, I wondered, an indicator of the onset of middle age and a consequent lack of adaptability? Or was there, as Aristotle suggested, something proper about bodies being in place, such that there was a relationship between being and implacement? In other words, was my concern wholly justifiable? However, in the midst of my feelings of vindication came a profoundly disturbing thought: Was I not a Christian theologian? How much "at home" was I supposed to feel in the world? Was the world really my place, or was my true destiny to be found somewhere else? Perhaps somewhere beyond the world?

Was "place" something to be cherished, or was it an idol, the purpose of which was to seduce me from my heavenly goal? As I watched the history of place unfold, I was able to discern at least three things. First, however much place was subverted by space and time—and place would contract to a mere point—there nevertheless remained an awareness that place could not be obliterated totally. Second, the traditions that most emphasized place were non-Christian and often non-religious. And third, the role of Christianity in the discussion of place was ambiguous and in need of urgent hermeneutical reconstruction if a sense of place was indeed foundational for ecological consciousness.

All I knew was that so far as I was concerned, place was important. Having a place, indeed having a sense of place, gave me a consciousness of belonging, and with that consciousness came the will and the desire to care for the place that gave me being-in-the-world.

In this chapter I hope to show that place is thoroughly ingredient in our self-understanding. Despite having undergone some hard times, the concept of place has endured and looks as though it may even be set to flourish. The urgency implicit in the chapter is that while place has been important for human self-understanding and correlatively for ecology, its journey within the Judeo-Christian tradition has not been without significant ambiguities. By claiming that in ecotheology we are faced with a "plea for place," I am suggesting that it is high time for a dialogue between Christian theology and place. I do not seek to solve the issue raised here, but rather to further focus the question.

Bereft of Place

"Can you imagine," asks Edward Casey, "what it would be like if there were no places in the world? None whatsoever! An utter, placeless void!" And he answers his own question: "I suspect that you will not succeed in this thought-experiment, which is not just difficult to perform (can you really eliminate any trace of place from your experience of *things*?) but also disturbing (can you really picture yourself in a *world* without places?)."[1] We take place for granted and it is really only when we make the effort to attend to our language and our thought that we realize just how much place-saturated our lives are.

Given that our lives are so place-oriented, it comes as no surprise to learn that we have a horror of displacement. "The emotional symptoms of placelessness—homesickness, disorientation, depression, desolation— mimic the phenomenon itself."[2] Separation from place has been made manifest in our era by the huge numbers of displaced persons. Yet, feeling

---

1. Edward Casey, *Getting Back into Place: Toward a Renewed Understanding of the Place-World* (Bloomington: Indiana University Press, 1993) ix.
2. Ibid., x.

as though we have no place can overcome us even in the circumstance of being in a place, perhaps, contrary to all outward impressions, in the right place. Here we experience not being "at home." We are creatures subject to the phenomenon of "place-panic." And to that end the philosopher Gaston Bachelard suggests that we may need to return, if not in actual fact, then in memory and imagination, to the very earliest places we have known.[3]

Perhaps that journey may curb our subsequent place-panic. Then again, perhaps not. Psychological though this phenomenon of place-panic certainly is, it is not restricted to the psychological sphere; it brings with it theological and philosophical consequences as well. Beneath them all is the dictum: Avoid the void at all costs. It is paradoxical, then, that much of the history of Western philosophy, science, and not a little theology has resulted in adopting the void—in the sense of infinite space—albeit with some unease.

This unease is reflected in Pascal's well-known cry that "the eternal silence of these infinite spaces terrifies me."[4] This is a cry that witnesses to the ascent of infinite space over place and consequently the displacement of place. Moreover, it is a "terror felt by such otherwise dispassionate figures as Descartes, Newton and Leibniz, each of whom attempted to deal with his phobia by papering over the abyss with philosophical and scientific speculation."[5] Closer in time, it is reflected to us in a more overtly theological way by Nietzsche, who, in speaking of the death of God, referred also to the accompanying horror of placelessness.

> What did we do when we unchained the earth from its sun? Where is it moving to now? Where are we moving to? Away from all the suns? Do we not continually stumble and fall? Backwards, sidewards, forwards, in all directions? Is there still an above and a below? Are we not wandering through an infinite nothingness? Do we not feel the breath of empty space?[6]

This sense of atopia, of being without place, has become part and parcel of our state of mind. This state of mind has unfortunate and potentially dangerous consequences with regard to ecology, for if we live with a sense of placelessness in our very own selves, shall we be too surprised if this very placelessness spills over into the environment and destabilizes by displacing?

3. Gaston Bachelard, *The Poetics of Space,* trans. Maria Jolas (Boston: Beacon Press, 1994).

4. Blaise Pascal, *Pensées* (London and New York: Everyman's Library, 1908) 61.

5. Casey, *Getting Back into Place,* xi.

6. Friedrich Nietzsche, *The Joyful Wisdom, Works* (London, 1910) 168, cited by Jürgen Moltmann, *God in Creation: An Ecological Doctrine of Creation,* trans. Margaret Kohl (London: SCM Press, 1985) 141.

In- and Out-of-Place

An ecotheology must, I think, move beyond the human even if the human being remains the point at which it begins and ends. It is inevitable that in arguing for the primacy of place, we shall find that we spend a lot of time discussing what place means for us, but this need not be problematic unless we regard place from a purely instrumentalist perspective, as did the early Heidegger.[7] Given that *oikos,* or "house," is embedded in ecology and economy, it follows that the natural world as conceived from an eco-perspective, far from being antithetical to dwelling-as-residing, invites it and makes it possible. But—and here we must diverge from an example of the early Heidegger—this does not mean that the earth is nothing but "the place where mortals—humans—dwell." As Casey insightfully points out,

> The earth's destiny is not to be the site for human habitation alone; nor does the idea of the earth as a house need to be conceived by analogy with a human house. It might well be the other way around: *our* dwelling places may draw their most telling inspiration from *its* dwelling places. This is certainly the case if the earth is indeed the primary place of residence for all species, the world's first housing agency, as it were.[8]

Thus by turning the house analogy on its head, we are given a way forward in our discussion of place.

It is a truism almost too obvious to be stated that a human being differs from a tree. These differences are manifest and many, but a primary difference of place often goes without comment. A tree simply stands in its own place. Indeed, its entire life is lived in one place, unless we allow for transplanting, which then, of course, is not effected by the tree. As Hans Jonas remarks, "With its adjacent surroundings, the plant forms one permanent context into which it is fully integrated, as the animal never can be in its environment."[9] The tree may well be displaced by external circumstance, but it will not displace itself, for it cannot move. The tree is in-place, and if it is to live, it cannot be other than in-place. To be out-of-place is death.

The case with animals, including, of course, human animals, is different. Not only *may* we change our place, we *have to* if we are to survive. Of

---

7. The early Heidegger of *Being and Time* pays only scant regard to place. The later Heidegger did attend much more to place, and at a seminar he gave at Le Thor in 1969 he stated that his thinking had traversed three periods, each with its own leading theme: Meaning, Truth, and Place. See Edward Casey, *The Fate of Place: A Philosophical History* (Berkeley: University of California Press, 1997) 244.

8. Casey, *Getting Back into Place,* 266.

9. Hans Jonas, "To Move and to Feel: On the Animal Soul," in *The Phenomenon of Life: Toward a Philosophical Biology* (Chicago: University of Chicago Press, 1966) 104.

all animals, human beings are among the most mobile. Casey points out correctly that "Appetite, along with memory and desire, calls for a continual change of place. . . . Changing places is not only a matter of tedium or restlessness. Our desires and interests, our very metabolism and musculature, conspire to keep us moving."[10] Therefore, changing place or getting out of place is a basic action that is part of our very nature. Freedom and mobility ensure that we, like other animals, will change places. But with such freedom comes the ever-present danger of being lost or misplaced. To curtail, at least to some extent, the danger of becoming lost animals, nonhuman as well as human animals have come to rely on territoriality as a means of maintaining a place that affords security and comfort. Mobility happens within a bounded place.

To be unplaced is a most frightening prospect for us. To admit that one has no place to go is to admit to desperate circumstance. Such an admission seems to lay one's very being open to question. It uncovers the formality of Heidegger's *Dasein* as Being-in-the-world and begs it to be clothed in terms of Being-in-place, and this because, as Gibson points out, "we do not live in 'space.'"[11] Most probably it is from this anthropological perspective that we perceive the threat of ecological devastation. "The earth is becoming uninhabitable," we cry. We fear, and justifiably, that no longer will it provide habitation or place in which not only we but all that exists may live. We know, however inchoately, that where we are—the place we occupy, however briefly—has everything to do with what and who we are, and finally, *that* we are.

Given the fundamental nature of place that has been adverted to in this section, we may wonder just why it was that place came to be not only neglected but actively suppressed. Why this chasm between the belittling of place on the level of official doctrine and the prominence of place in ordinary life? In an attempt to answer this, we need to recount a summary history of place.

Aristotle

It is Aristotle who maintains that *where* a thing is constitutes a basic metaphysical category.[12] He also claims that place has some power, an

10. Casey, *Getting Back into Place*, xii.

11. J. J. Gibson, *The Ecological Approach to Visual Perception* (Hillsdale, N.J.: Erlbaum, 1986).

12. J. Barnes, *The Complete Works of Aristotle* (Oxford: Princeton University Press, 1984) 5th rev. ed., Oxford translation, *Categories*, 2a1, "of being-in-a-position." See also *Categories*, 5a9-14: "Place, again, is one of the continuous quantities. For the parts of a body occupy some place, and they join together at a common boundary."

assertion that remains somewhat opaque at this stage but will become clearer as we see just what part *place* plays in Aristotle's system.

As a preliminary definition of place, Aristotle claims that just as a vessel such as a glass or jug surrounds its content—say, air or water—so place surrounds the body or group of bodies located within it.[13] That means that the purpose of place is to contain. But he immediately recognizes that unlike a vessel, which can be moved from place to place, place itself cannot be moved. It is bodies that move from place to place, not the places themselves. That leads him to emend his first definition of place by adding the rider that a place cannot itself be changing or moving: it must be unchangeable. Thus he moves to his final definition: place is "the first unchangeable limit of that which surrounds."[14]

Place, according to Aristotle, always contains and surrounds the body, and yet it is not, as it is for Plato, simply a receptacle. Place is "actively circumambient."[15] By containing and surrounding, place gives shape and definition to physical bodies. Or, as Casey puts it, "the limiting power is *already in place*."[16] Limit is ingredient in place from the beginning—in fact, *as* the beginning of an ordered natural world—and is not imposed by an external ordering agent. Rather, places have their own independent potency, all of which means that place has some power, as Aristotle claims.[17]

The world that Aristotle bequeaths to us is a snug world, a cozy world. The world, he suggests, is filled to the brim with snugly fitting, proper places. He claims that "it is obvious that one has to grant priority to place."[18] He demonstrates this by showing that place, beyond providing mere position, gives bountiful aegis—active, protective support—to what it locates.[19] In Aristotle's hands, place certainly does exert a certain influence: it has the power to make things be *somewhere* and to hold and guard them once they are there. As Casey correctly notes, "Without place, things would not only fail to be located; they would not even be *things:* they would have *no place to be the things they are*. The loss would be ontological."[20]

13. Aristotle, *Physics,* 209b3-6: "Now if place is what primarily contains each body, it would be a limit, so that the place would be the form or shape of each body by which the magnitude or the matter of the magnitude is defined; for this is the limit of each body."

14. Aristotle, *Physics,* 212a20-21: "Place . . . is rather what is motionless . . . the place of a thing is the innermost boundary of what contains it."

15. Casey, *The Fate of Place,* 55.

16. Ibid. Emphasis in the original.

17. Aristotle, *Physics,* 208b11: "The locomotions of the elementary natural bodies—namely, fire, earth, and the like—show not only that place is something, but also that it exerts a certain influence."

18. Casey, *The Fate of Place,* 71.

19. Ibid.

20. Ibid. Emphasis in the original.

Disillusionment with Place and Reenchantment

The dominant reason for the discontent with the Aristotelian system of place is the very snugness adverted to previously. Against this closed Aristotelian world, albeit a cozy one, is juxtaposed an emphasis that is directed toward space, where space is understood as something undelimited and open-ended. "While place solicits questions of limit and boundary, and of location and surrounding, space sets these questions aside in favor of a concern with the absolute and the infinite, the immense and the indefinitely extended. . . . In this unequal battle, spacing-out triumphs over placing-in."[21]

In this transition from place to space, we move from the naturalistic world of Aristotle's system, "in which the vernacularity of place, its habitability and idiosyncrasy is predicatably prominent, to a theological *Weltanschauung* in which the infinity of space becomes a primary preoccupation."[22] Part of the challenge for an ecotheology must be to trace at least some of the impact of the trajectory of theology upon the story of place. Here it must be admitted that the part that theology has played in the transition from place to space is ambiguous.

Christian theology sought to explore, develop, and especially defend the issue of divine immensity. Much of this history must be skipped over in silence, but in outline the argument unfolds as follows.[23] If God is limitless in power—and mainstream Christian thought held this line consistently—then God's presence in the universe at large must also be unlimited. With the rise of modern thought, Aquinas's understanding of God as *in omnibus rebus et intime* ("in all things and intimately") was subverted. Instead, God was reduced to natural categories; hence the analogy of divine ubiquity and spatial infinity. Thus historically, if not logically, divine ubiquity entails spatial infinity. It further follows that the physical universe itself must be unlimited if it is to be the setting for God's ubiquity as well as the result of divine creation. Thus the particular, the local, and the bounded gave way to the delimited and the boundless. Casey concludes: "It comes as no surprise to learn that the increasing hegemony of Christianity supported both forms of infinity: that of God as the ultimate monotheistic being and that of God's universe as the ultimate monothetic entity."[24]

It is by now a commonplace to read how the roots of modern science are to be discerned from a religious background, specifically the Judeo-Christian background. The theological background and assumptions of the infinity of God—a Christian concern indebted to Greek philosophy—set

21. Ibid., 77.
22. Ibid.
23. See ibid., 79–150.
24. Ibid., 77.

the stage for a comparable concern with the spatial infinity of the physical universe by the natural scientists and philosophers who began to mathematize nature in the sixteenth and seventeenth centuries. This resecularization of the world by way of quantification would not have been possible without the theological reflections of the preceding centuries. Theology and physics were closely allied in their common effort to conceive of space in utterly maximalist terms.

The line between exploration and colonialist expansion was always thin and indistinct, leading Casey to pose the following question: "If theology, especially Christian theology, is universalist in its aims, why should not the new physics—standing on the shoulders of this ambitious theology—proclaim truths that hold for every material object in the universe?"[25] In fact, we see this colonizing tendency at work in the attempts of Cartesian and Newtonian physics to appropriate whole realms of reality under their control. And in the case of Christian theology, as well as the approach of modern science, we see that the power of place was subverted. So it was that by the end of the eighteenth century, the idea of universal space came to be regarded as obtaining not just for the external world and for God but also for the mind of the knowing subject.[26]

With the work of Kant in the eighteenth century, we have reached the stage where place is compressed to a mere point. Yet, Janus-like, Kant not only completed the subversion of place but he also anticipated the reinvigoration of place by suggesting that it is from the human body that we trace spatial dimension. He discovered that the body plays a fundamental role in spatial dimensionality.[27] True, this discovery lay unheeded until the twentieth century, when it was rediscovered by Alfred North Whitehead and further developed by Maurice Merleau-Ponty.

---

25. Ibid.

26. Although it seems as though place and space must refer to things outside the body, the reality turns out to be rather more subtle. Beginning with Descartes there is the subsumption of every sensible appearance under a representation whose status is mental. For any appearance to be apprehended, it must assume the format of a representation, the sum total of which makes up Mind itself. This representationalism takes in everything in the universe, including space and time. In the twentieth century the phenomenological recovery of place will ensure that place is considered from the perspective of the inner life of mind (Bachelard), as well as the exterior of the body (Merleau-Ponty).

27. "Directionality such as 'right,' 'left,' 'up,' 'down,' 'back,' 'front,' . . . these paired terms, taken together, describe the three dimensions of space: the dimensionality of space follows from the dimensionality of the body. . . These are not oriented in and by themselves; they require our intervention to *become* oriented. Nor are they oriented by a purely mental operation: the *a priori* of orientation belongs to the body, not the mind." Casey, *The Fate of Place*, 187–210.

Whitehead, arguing from a process perspective, criticized the seventeenth-century thinkers for their view of space and time as one of "simple location." Simple location was the belief that anything in space was nothing other than a piece of matter, without any essential reference to the relations of that bit of matter to other regions of space and to other durations of time. Simple location entailed the assumption that matter is non-relational. In contrast to this, Whitehead opted for a relational understanding of reality. That permitted him to argue for the primacy of body in relation to place. For Whitehead, the body is not so much a thing as an "event," and as such it is not a receptacle for sensations but an active participant in the scene of perception.[28] The body prehends place, and it does so by virtue of its "withness." "We see the contemporary chair, but we see it *with* our eyes; and we touch the contemporary chair, but we touch it *with* our hands."[29] Not only do we experience the chair with our body but also the place of the chair. "Place . . . arises *within the withness* essential to the body's primitive prehensions . . . of its environing world."[30]

Whitehead understands the body to be radically im-placed and maintains that it is because of this bodily im-placement that we are enabled to recover a sense of place. Simple location (the atomistic view) gives way to modal location (the relational perspective). We are always in place because we are embodied, and all that is embodied is in place.

The process of bodily implacement was further developed by the work of the phenomenologist Maurice Merleau-Ponty. He maintained that the body is "our general medium for having a world."[31] It is through our bodies that we encounter the world, and the lived body has its own "corporeal

28. Alfred Whitehead, *Science and the Modern World* (New York: Cambridge University Press, 1926) 73. See also his *Process and Reality: An Essay in Cosmology* (London: Cambridge University Press, 1929) 111f.: "An event is a nexus of actual occasions interrelated in some determinate fashion in some extensive quantum. . . . For example, a molecule in a historic route of actual occasions; and such a route is an event."

29. See *Process and Reality,* 89. Whitehead points out that this notion of "withness that makes the body the starting point for our knowledge of the circumambient world," 112.

30. Casey, *The Fate of Place,* 215. "In contrast to Newton's view that nature is merely, and completely, *there,* externally designed and obedient, on the Whiteheadian model the body is the arena in which the here and the there conjoin inextricably . . . The body, or more exactly *my own body,* is unique in bringing together here and there in a manner that resists the allure of simple location, according to which the 'here' is merely the pinpointed position of my body regarded as an indifferent thing and the 'there' the equally pinpointed spot of the contemporary object opposite me. Instead, the 'there' *ingresses* into the 'here' and vice versa."

31. Maurice Merleau-Ponty, *Phenomenology of Perception,* trans. Colin Smith (London: Routlege & Kegan Paul, 1972) 146.

intentionality."[32] This intentionality of the body is what binds us to the life-world we inhabit. It is not just the body but the movement of the body that is productive of space.[33] As Casey succinctly records, "Galileo's apothegm 'It moves!' is superseded by Merleau-Ponty's operative dictum 'I move.'"[34] It is the motility of the body that exhibits what Merleau-Ponty calls "expressive movement." Indeed, it is through the motility of our body that we are continually orienting ourselves in the place where we find ourselves to be. We are not simply in space and time, we inhabit them. Thus he suggests that we think in terms of belonging to space and time. "My body combines with them and includes them."[35] Here we have moved far beyond the Aristotelian notion of place as that which surrounds to one of place as mutual penetration. But if place merges with body, what can be meant by moving from place to place? Where does place end and body begin?

Doubtless there is a subjective edge to Merleau-Ponty's understanding of place. A place is not merely something I inhabit with my physical body. A place is also somewhere that I, while not being physically present, nevertheless inhabit by way of its familiarity and my own corporeal habituality. Places are experienced and known not only by actually being there now but also through the customary body and its habitual activity that imprints itself upon the psyche. The subjectivist emphasis is firmly held in check by his insistence that place comes to be known only through the body. To be sure, this body is a phenomenal as well as a physical body, but the key is "as well as" and not "instead of." And by way of body we see that the disillusionment with place begins to yield to a reenchantment with place.

Ancient Wisdom from My Place

As I sit at my computer, the occasional car drones down the highway. Looking around, I see an expanse of houses, manicured lawns, paved footpaths, and bitumen roads. A lot of Australia looks just like this. But it wasn't always this way. No, there was a time when there were no bitumen roads or concrete driveways. No rabbits or sheep. No oaks or elms in our parks. It was a very long time ago. It was when the only people here were the indigenous people. It was a time, so their descendants tell us, when "place" was everything.

---

32. Casey, *The Fate of Place,* 229, and "How to Get from Space to Place in a Fairly Short Stretch of Time: Phenomenological Prolegomena," in *Senses of Place,* ed. Steven Feld and Keith H. Basso (Santa Fe, N.M.: School of American Research Press, 1996).

33. Merleau-Ponty, *Phenomenology of Perception,* 387.

34. Casey, *The Fate of Place,* 230.

35. Merleau-Ponty, *Phenomenology of Perception,* 139–140.

Although place may be fundamental to the way we feel, it is not fundamental to the way we think. That is what gives plausibility to Casey's claim that place is almost too familiar, almost too close at hand. Nevertheless, as the previous section has made clear, we are able to discuss what we mean by place, what place means to us, and, most significantly, how place has evolved and changed throughout history. Indeed, it is the vantage point of history that provides one of the platforms for the discussion of place. But how would things be without history?

Merleau-Ponty raised the question of whether it is possible to conceive of traditions in which our concept of history is simply absent.[36] Tony Swain has suggested that we may answer this question in the affirmative. The earliest intimations of ontology that we can discern from indigenous Australians points to a tradition based upon place rather than history. Swain refers to this "history of Australian Aboriginal being" as a "hermeneutics of ubiety,"[37] that is, a hermeneutics of whereness or of being in a definite place.[38]

The importance of place in indigenous Australian consciousness has made its presence felt in various disputes about land over recent years. Yet, insofar as most people persist in interpreting the land from the perspective of ownership, the fundamental ontology of place that Swain is seeking to discern remains invisible. The battle lines, such as they are, are drawn out on the plane of history. But although indigeneous Australians have a concept of history, it is not the same as the dominant Western idea of history as a linear progression. And it is this recognition of differing understandings of history that has sometimes led to the mistaken belief that the ontology of the indigeneous peoples is timeless. Instead, their ontology is related to place and can therefore be understood only as a hermeneutics of ubiety.

According to Swain, there is a recognition of the rhythms of life; however, these rhythms are recognized because they are implaced. Nothing exists beyond the events themselves. There is no space into which the events fit. Instead, there are simply abiding events residing in a place.[39]

The issue of the relation of a historical consciousness undergirding indigenous Australian culture has been obscured by the term "Dreamtime." Indeed, so problematic has the term become that it is now more usual and more accurate to refer to "the dreaming." Certainly the word

---

36. Tony Swain, *A Place for Strangers: Towards a History of Australian Aboriginal Being* (Cambridge: Cambridge University Press, 1993) 4, citing M. Merleau-Ponty, "Phenomenology and the Sciences of Man," in *Phenomenology and the Social Sciences*, ed. M. Natanson (Evanston, Ill.: Northwestern University Press, 1973) 1:103.

37. Swain, *A Place for Strangers*, 4.

38. Definitions for "ubiety" are taken from the Oxford English Dictionary.

39. Swain, *A Place for Strangers*, 15–19.

witnesses to something fundamental in the indigenous consciousness; however, the concept behind the Dreamtime is not temporal as *time* would indicate but spatial.[40] Although the past, future, and present are recognized, they are not worked into an ontology. Rather, what we find is that abiding events and rhythmic life events are cojoined through place.

There is no notion of space in the singular.[41] The only unit of cosmic structure is the place. Not surprisingly, this means that there is no utopia. It also means that there is no creation.[42] Instead, one may speak only of independent place-shaping events. The world is not made. But worlds take shape.

Even consciousness develops from place.[43] Thus the ancestral place is not simply the place where one was born (with all the importance that connotes); it is also the way in which a person's core identity is determined. This determination of core identity implied that people were, in some difficult-to-understand but nevertheless real sense, extensions of place. Indeed, so fundamental was this belief among some Australian indigenous communities that the emphasis was put upon the place of one's birth when speaking about the derivation of one's rights and obligations. Even one's body and one's paternity derived their importance from place. That is not to say that kinship and the body are unimportant; instead, it is an affirmation that their importance is derivative from the primacy of place.

Another example of this primacy of place comes the tradition of some indigenous males who maintain that children are not born of women's bodies—this does not imply a misunderstanding of biological birth, as a facile response might erroneously assume—but of a place. Perhaps that is the reason that lies behind the custom of immediately placing the newborn child in a small earthy depression, from which it is then "born"—an act stating that the child comes not from the mother but from a location or place.[44]

With increasing outside contact beginning first with the Melanesians, then with the Indonesians, and finally with white invasion, the indigenous peoples' cosmology and ontology began to change. Personhood defined as an extension of place gradually moved to a nostalgia for life after death in a utopian realm.[45] Place became subverted. And here we are reminded that colonialist expansion and theology made their mark once more. True, the stories continue to be told, but they are now spoken into a vat of history,

40. Ibid., 21–22.
41. Ibid., 29.
42. Ibid.
43. Ibid., 33.
44. Ibid., 44.
45. Ibid., 287.

which means they are difficult to hear. Perhaps they are impossible for us to hear fully. Perhaps all we now discern is the trace of place.

God and (as)? Place

When people talk of creation, their first emphasis usually falls upon the some*when* rather than the some*where*. However, admit the priority of the second question and the way is immediately open to place. Where does creation take place? We tend not to think of the scene of creation. "But if we ask ourselves not *what* was created or even *how* it was created but *where* creation occurred, we realize that it could not have happened just anywhere, much less nowhere."[46] Dhammai legend records that "at first there was neither Earth nor Sky. Shuzanghu and his wife Zumiang-Nui lived above. One day Shuzanghu said to his wife, 'How long must we live without a place to rest our feet?'"[47] As a rider we may ask, "Above *where?*" The very language that tells the story of the making of place is itself placial. Cosmogonic narratives are not only recountings of events in time but also of things in place.[48] If it is true that in the beginning was the Word, it is no less true that in the beginning was the Place. The dictum that *ex nihilo nihil fit* ("from nothing, nothing comes") holds true, for all that comes to be comes from place and dwells in place.

To suggest that in the beginning was the Place is to posit an ultimacy for place in such a way that we may speak of God being place. Admittedly, the notion of God as place is drawn from the rather idiosyncratic kabbalist tradition that envisions God as MAKOM, or place.[49] According to Philo,

> God himself is called a place, by reason of His containing things, and being contained by nothing whatever . . . . for He is that which He Himself has occupied, and naught enclosed Him but Himself. I, mark you, am not a

46. Casey, *Getting Back into Place*, 18.
47. Casey, *The Fate of Place*, 3.
48. Ibid., 7. "Events . . . call for cosmic implacement: no event can happen unplaced, suspended in a placeless *aither*. This includes the event of creation itself. It, too, must have its place. Integral to cosmic creation is the creation not just of places for created things as such but a place for creation (and thus for the creator)."
49. Moltmann, *God in Creation*, 154: "There is also a Jewish-kabbalistic tradition, according to which one of the names for God is MAKOM, which means the infinite, pure and primary place." The term "kabbalist" is used to refer to Jewish mysticism from the twelfth century; see Moshe Idel, "Qabbalah," *The Encyclopaedia of Religion*, ed. Mircea Eliade, vol. 12 (New York: Macmillan, 1987). It can also be interpreted in a wider sense, signifying all the successive esoteric movements in Judaism that evolved from the end of the period of the Second Temple and became active factors in the history of Israel; see "Kabbalah," *Encyclopaedia Judaica*, vol. 10 (Jerusalem: Keter Publishing House, 1972).

place, but in a place; and each thing likewise that exists . . . and the Deity, being contained by nothing, is of necessity Itself Its own place.[50]

Philo is here relating to the tradition of God as MAKOM, or place, and thus buying into a usage that was prevalent before his time. A midrashic text on Genesis 8:10 claims that "God is the place of the world and the world is not his place."[51] It is likely, of course, that originally the term "place" was used only as an abbreviation for "holy place," the place of the Shekinah. It is also likely that as Jewish theology became more abstract over time, "place" was understood as a cipher for divine omnipresence, and thus almost the opposite of place.

That God is Place never figured prominently in mainstream Jewish thought, nor, it must be said, in Christian speculations, despite the opportunity afforded by the incarnation. Instead, mainstream Christian theology sought to speculate about divine immensity. That is curious, since on the surface it seems that the doctrine of the incarnation would be especially suited here.[52] The descent of divinity into the world of materiality and corporeality would seem to suggest ready connections between place and God. More significant, as Geoffrey Lilburne points out, is the fact that ecotheology is seeking to redress a world already in the thrall of despoliation, and therefore we are confronted by weighted choices rather than equally balanced alternatives. All this implies the need for a redemptive as well as a creative presence of God.[53] The incarnate presence of Christ is suited to this task, speaking as it does of redemption and transformation.

Lilburne presents his own interpretation of the interrelationship between the macro and micro aspects of ecotheology.[54] He derives this "dialectic," as he calls it, from a prior investigation into the locality of the kingdom of God, from which he discerns that "the kingdom of God has both a primary local meaning and a secondary universal meaning."[55] The primary

50. Cited in "Space and Place," *Encylopaedia Judaica,* vol. 15, from Philo's treatise *De somnis.*

51. Ibid. This midrash was later picked up by Nahmanides (1194–1270), the Spanish philosopher, kabbalist, and biblical exegete, who stated that in his opinion the sages meant by this dictum that God is the form of the world, since form is the realisation (the entelechy) of the perfection of what is contained in the world and is also its limit, since it prevents the spreading out of the world's dimensions beyond its form. The notion of limit and boundary, together with the dynamism implicit in entelechy, is reminiscent of Aristotle's notion of place, albeit with qualifications.

52. Duncan Reid's essay in Chapter 4 of this volume is an example of the incarnation being used in the service of ecotheological principles.

53. Geoffrey R. Lilburne, *A Sense of Place: A Christian Theology of the Land* (Nashville: Abingdon Press, 1989) 90.

54. Ibid., 102.

55. Ibid.

emphasis on the local meaning is to be found in Jesus' practice of table fellowship. There is a close connection between Lilburne's notion of primary local meaning and the secondary universal meaning, as compared with Bachelard's discovery of "intimate immensity," where the larger world is experienced in miniature.[56] For both writers there is a dialectic between place and space, and for both it is from the perspective of place that space is encountered. Bachelard begins this journey in memory, while Lilburne begins from within the Christian community. Lilburne suggests that the place of God's presence is in the Christian community, and this in turn implies that "the Incarnation must be understood in terms of an ongoing process."[57]

Between the polarities of God as Place and God as incarnate presence in the Christian community can be situated the relationship between God and place as it is variously portrayed in the Scriptures. And it is this scriptural witness that is so fascinating, not least because of its ambiguous stance toward place. On the one hand, places and place names in the Bible carry a strong resonance with them. Consider, for example, in the Hebrew Scriptures, Egypt, Sinai, Canaan, Shechem, Mamre, Mount Zion, Jerusalem, and Babylon; while the New Testament speaks of Bethlehem, Nazareth, the Sea of Galilee, the Jordan River, Bethany, Jerusalem, and Golgotha. These places reach out to us across the centuries and are much more than mere names. Indeed, Burton-Christie reminds us that "they carry with them and call forth a web of narratives and events which are rooted in particular places."

The biblical story can in a sense be told in terms of the texture and character of these places.[58] For both the Old and New Testaments, place is appreciated concretely and symbolically.[59] Place anchors the biblical story, from the garden, to the mountain, to the holy city of Jerusalem.[60] Each of

56. Bachelard, *The Poetics of Space,* chaps. 7 and 8. Instead of the cosmological claim that the world is a house, Bachelard's topoanalysis suggests that the house is a world. And to this house we may return in memory. Indeed, it is in memory that we experience the fascinating phenomenon of intimate immensity, which means "to experience oneself in a larger world in miniature."

57. Lilburne, *A Sense of Place,* 104–105. "If in his life Jesus sanctified places by his presence and loving relationship, so too the community was to sanctify and care for the places of their life. If God really was present where the name of Jesus was spoken, if the Spirit mediated the presence of the risen Christ to all Christ's people, then the Incarnation was an ongoing process that needed to be a central element of the understanding and practice of the Christian communities . . . they must seek to sanctify the places of their life in the name of Jesus. According to this extension of the Incarnation, in the localities of their lives God is present."

58. Douglas Burton-Christie, "Living Between Two Worlds," *Anglican Theological Review* 79 (Summer 1997) 420–421.

59. Ibid., 423ff. for examples.

60. Ibid., 422.

these places becomes revelatory, laden with meaning and promise. Indeed, the very notion of promise itself has connections with place, as is made apparent by the lament of the exiles in Psalm 137. How could they sing the Lord's song in a foreign land or place? The agony of prayer cannot be considered separately from the reality of their dislocation.

Understood from this perspective, it might seem that the Bible has an unambiguous regard for place. However, there is another disposition toward place with which it must be juxtaposed. There is the tradition of exile and wandering, a situation of having no-place. There is a certain suspicion of place—that one might seek for too much comfort and security from place. There is a fear that however sacred a place may be, however soaked by divine presence, it may come to be seen as a substitute for right relationship with God. In the Hebrew Scriptures we find a theology of exile and diaspora set alongside a theology of place. And it is significant, claims Burton-Christie, that "the symbolic center of Jewish identity that emerges from within the exile experience . . . is the scroll. This effectively liberated Jews from their dependence upon place and created the possibility of multiple . . . centers throughout the diaspora."[61]

In the New Testament, too, we see a deep suspicion of place. For example, Jesus is an itinerant preacher moving from place to place. He has nowhere "to rest his head" and evinces a certain uneasiness about being at home and claims that a prophet is not recognized in his or her own country. Mount Gerizim and Jerusalem both give way to the true worship, which will be carried out "in spirit and in truth"—significantly unplaced, or at least not related to any particular place. The Christians at Philippi are assured that their "true homeland is in heaven" (Phil 3:20). Similarly, the Letter to the Hebrews claims that "here we have no lasting city, but we seek the city that is to come" (Heb 13:14). Evidence could be multiplied, but it is clear that the Bible portrays a tension between place and journey, and "there is a deep and abiding ambivalence about where and how God is to be encountered."[62]

Conclusion

If nothing else, I hope to have shown that place is a potent reality. Throughout its history and despite all attempts at subversion, it has survived and, from time to time, even flourished. It is integral to personhood, indeed in some cultures constitutive of ontology. Yet in suggesting ecotheology as "a plea for place," we immediately make connections between God and

61. Ibid., 425.
62. Ibid., 422.

place, thereby disclosing deep ambiguities. These ambiguities are adverted to in the Scriptures themselves as well as in the historical impact of theology on the "fate of place." More significantly, these factors operate not only on the plane of abstract ideas; they also impact at an existential level, prompting us to question how much at home we should feel in the world.

As I mentioned in the introduction, my purpose in this chapter was to further focus the question of place in ecotheology. This I have done by pointing to the rather uncomfortable fit between the sense of place and its indispensable nature which we hold at an existential level and the ambivalent impact of Christian theology upon the concept of place. Perhaps this ambivalence cannot finally be resolved, but certainly it is a tension that deserves further study.

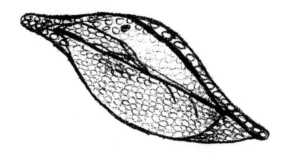

Both-Way Mussel

Him live in saltwater
And in the fresh
The both-way mussel.

We women come
And walk round in the mud.
We feel . . . , we feel . . . ,
We feel the lumpy one!
We pick 'em up
And put 'em in the bag.
Still we're walking round
In the one place where they live.
We're circling in one place.
We feel, we pull and pick 'em up.
And when the bag is full
We go where we left our things
And boil tea.
We throw the mussels in the fire.
They open up,
We eat 'em.
They're lovely,
They're beautiful!

And see . . .
Sometimes we go
To get the both-way mussel
And sometimes we don't.

Brenda Alimankinni
Tiwi Islands

Chapter 11

Situating Humanity

Theological Anthropology in the Context of the Ecological Crisis

James McEvoy

Christian faith makes an immense claim: that in the life, death, and resurrection of Jesus of Nazareth, God is definitively revealed, and this revelation makes sense of human life and the cosmos.[1] This is not the subjectivist claim that a person's faith simply changes his or her perspective on the world, and hence the way in which he or she engages in the events of life. Rather, Christians believe that the life of God is made present in the world in the person of Jesus, and that this divine presence illumines the human situation.

Karl Rahner's theology is a fine example of sustained reflection on the divine Word illumining the human situation. A central theme of Rahner's theological project is the inner relationship between the question that is human existence and the revelation of God in the person of Jesus.[2] In an essay on the human search for meaning written toward the end of his life,

1. Mark Worthing studies this universalist claim in the thought of two prominent twentieth-century theologians, Karl Rahner and Wolfhart Pannenberg, in *Foundations and Functions of Theology as a Universal Science: Theological Method and Apologetic Praxis in Wolfhart Pannenberg and Karl Rahner,* European University Studies: Series 23, Theology; vol. 576 (Frankfurt am Main: Peter Lang, 1996).

2. See Karl Rahner, *Foundations of Christian Faith: An Introduction to the Idea of Christianity,* trans. William V. Dych (London: Darton, Longman & Todd, 1978). Rahner regarded this work as a summary of his theological system. For an articulation of the inner relationship between human existence and Christian revelation in his later work, see Rahner, "The Question of Meaning as a Question of God" and "Jesus Christ the Meaning of Life," *Theological Investigations* 21: *Science and Christian Faith,* trans. Hugh M. Riley (London: Darton, Longman & Todd, 1988) 196–207, 208–219.

Rahner concludes that "the question about an absolute meaning, if it is really accepted and if it is permitted to articulate itself in its ultimate form, yields of itself the existence of an absolute meaning as something real and thus the existence of God himself."[3] In the context of this understanding of the human, Rahner writes about Jesus as the self-expression of God—the answer to humanity's question about absolute meaning.[4]

At the end of the twentieth century, the ecological crisis stands out as a situation in need of illumination. The threat to forests, rivers, seas, and air is well documented.[5] And if Christianity's claim about the importance of the divine is to be credible, theologians must explore the way in which the divine presence illumines the ecological crisis. The question arises: How is God's presence salvific when the earth is under threat? Theologians approach this question in different ways. In *Jesus the Wisdom of God: An Ecological Theology,* Denis Edwards argues that there is a "profound inner link between an adequate theology and an ecological stance."[6] He argues that theology is necessarily ecological. His argument moves from the nature of theology to its ecological consequences. Edwards delves deeply into the Christian tradition and says that if the Trinity, creation, and humanity are to be understood adequately, they must be understood in terms of the key concept of relationships. He argues that adequate understandings of God as Trinity, of creation, and of humanity provide a theological basis to ground the intuition of many Christians that theology necessarily involves ecology. Edwards expresses the hope that his exploration "may open up a way of healing for our Earth and its community of life."[7]

In this chapter I hope to explore further dimensions of this profound inner link between theology and ecology and show that theology has an important contribution to make toward a solution to the ecological crisis. Not only does reflection on the nature of theology articulate an inner link between an adequate theology and an ecological stance. I hope to show, even if only in outline, that when the root causes of the ecological crisis are considered, a relationship with the God of Jesus Christ sheds a great deal of light—it overcomes the modern instrumentalist approach to nature. And more than simply overcoming this problem, I hope to show that a solution which deals with the roots of the environmental crisis *requires* such a relationship with the divine.

3. Rahner, "The Question of Meaning," 206.

4. See Rahner, *Foundations of Christian Faith,* 212–228.

5. For a passionate account of the threat to the earth, see Larry L. Rasmussen, *Earth Community, Earth Ethics,* Ecology and Justice Series (Maryknoll, N.Y.: Orbis Books, 1996).

6. Denis Edwards, *Jesus the Wisdom of God: An Ecological Theology* (Homebush, N.S.W.: St. Pauls, 1995) 2.

7. Ibid.

The first section of this chapter examines the causes of the ecological crisis. Here I am not attempting to identify all the precipitating conditions of the crisis in order to provide a fully fleshed-out historical explanation. Such an account is well beyond the scope of this chapter. Rather, using Charles Taylor's hermeneutical study of the modern identity, *Sources of the Self: The Making of the Modern Identity*,[8] I hope to identify the key understandings that inform prevailing attitudes to nature and therefore govern humanity's relationship with creation.

In *Sources of the Self*, Taylor describes an increasingly instrumentalist approach to nature in which nature is viewed as a neutral domain to be understood in order to be mastered, and people are regarded as free when they are shaping themselves over against reality. Taylor traces the rise of this approach to nature from its roots in the work of Enlightenment thinkers like Descartes and Locke to its "flowering" in the ecological devastation of our time. It is Taylor's argument that this instrumentalist understanding shapes the practices and institutions of our time. He argues that these understandings are significant and that a full account of historical causation must give a central place to the self-understandings that inform contemporary practices.[9]

Where, then, should we look for a solution to the ecological crisis? In the second section of this chapter, I argue that we need to look beyond the instrumentalist understanding. A move beyond the instrumentalist stance requires a renewed and deeper appreciation of creation. Such a move also demands a renewed understanding of human agency, an understanding of humanity in relationship with the natural world and recognition of the demands that the natural world makes on us. I want to argue further that this renewed view of personhood entails an understanding of the person in relationship to what is ultimately good—what Taylor calls a constitutive good or moral source, the love of which enables us to do and be good. He articulates these understandings in the first part of *Sources of the Self* and elsewhere.[10] Taylor's picture of the modern identity is not uncontested; he

8. Charles Taylor, *Sources of the Self: The Making of the Modern Identity* (Cambridge, U.K.: Cambridge University Press, 1989).

9. For Taylor's view of the relationship between ideas and their associated practices and institutions, see "A Digression on Historical Explanation," in *Sources of the Self*, 199–207 and 306–314. For a similar view of the relationship between ideas and practice, although in a different context, see Michael J. Sandel, *Democracy's Discontent: America in Search of a Public Philosophy* (Cambridge, Mass.: Belknap Press, 1996).

10. See Taylor, "The Moral Topography of the Self," in *Hermeneutics and Psychological Theory: Interpretive Perspectives on Personality, Psychotherapy, and Psychopathology*, ed. Stanley B. Messer, Louis A. Sass, and Robert L. Woolfolk (New Brunswick, N.J.: Rutgers University Press, 1988) 298–320; Taylor, "The Dialogical Self," in *Rethinking Knowledge: Reflections Across the Disciplines,* ed. Robert F. Goodman and Walter R. Fisher (Albany, N.Y.: State University of New York Press, 1995) 57–66.

writes in dialogue with other major thinkers about the nature of modernity, particularly Michel Foucault, Jürgen Habermas, and Alasdair MacIntyre. Space does not permit me to rehearse their perspectives on modernity and human agency but only to point toward the argument which, in my view, gives the best account.[11]

In the midst of the ecological crisis, Taylor's views of both the modern identity and human agency challenge us moderns to a more coherent understanding of ourselves as persons in the world, an understanding that could help to overcome the crisis. We are challenged to grasp the *situated* nature of human existence and to recognize that we only come to know ourselves already situated within the natural and social environment, and in *relationship* to a moral source. In the final section of this chapter I want to outline key dimensions of Rahner's theology of God and the human person. I argue that his theology of humanity in the presence of absolute Mystery adequately accounts for our embeddedness in the social and natural environment and in a relationship to God.

Instrumentalism and the Modern Identity

How is it that the earth came to be under such threat? Is it that we have built too many automobiles, chainsaws, bulldozers, and smoke-belching factories? These products of industrial society certainly affect our environment. Changes in industrial, technological, and administrative arrangements over the past few hundred years have significantly altered the face of the earth. But an analysis of the ecological crisis that centers on external arrangements is a redescription of the problem rather than an explanation. Taylor argues that we come closer to an adequate explanation by examining how people today can treat forests, mountains, and oceans as standing reserve for human use. He holds that a very influential strand of thought fostered by the Enlightenment objectifies nature, seeing it as neutral and to be shaped according to human purposes.[12] It is Taylor's understanding of the Enlightenment view of nature and human agency that I now want to outline.

According to Taylor, eighteenth-century Deism played a major role in preparing for the Enlightenment. Deist thinkers such as Frances Hutcheson and Matthew Tindal saw the world as a providential order. They held that God's goodness and benevolence were expressed in a world in which the purposes of different creatures interlock. The life of each creature serves

11. Taylor engages with the thought of Foucault, Habermas, and MacIntyre particularly in Parts 1 and 5 of *Sources of the Self*.

12. In the ensuing paragraphs I am following Taylor's line of argument in chapter 19 of *Sources of the Self*, 321–354.

the entire order. In this worldview, what makes a proper human way of life good is living according to the design of nature. For Enlightenment thinkers, however, the Deist view of the world as a providential order—as designed for the best—was difficult to believe, and for two reasons. First, it failed to take suffering and disaster seriously, as portrayed in Voltaire's *Candide*. Second, its picture of a world where virtue and self-interest come so neatly together lacked credibility.

Taylor agues that in turning away from the Deist vision of the order of things, Enlightenment thinkers developed a very different picture of both the human person and nature. In the dominant Enlightenment worldview, individuals no longer looked to an external order to understand themselves but rather looked inward. They discovered their purposes within. Each person is seen to have an internal purpose with no *a priori* allegiance to a pre-existing order. This reenvisioning of the human person took place around the key concepts of reason and freedom. The task of reason was no longer seen as that of locating entities within the providential order; rather, the stance of disengagement became critical to reasoning: the human person stood over reality in order to understand and master it. Human freedom was no longer defined in relation to the whole—the *polis* or the republic—as in ancient conceptions. An atomist concept of freedom developed in which the subject was seen to enjoy natural liberty in the state of nature. With the vision of the order of things dismissed, creation or nature was also conceived differently. Nature was seen as a neutral domain, which humanity would explore in order to understand and master it. Because nature is seen as neutral and human agency as disengaged, the Enlightenment view is called instrumentalist.

Taylor argues that alongside the concepts of reason and freedom, a moral motivation is crucial to an accurate understanding of the Enlightenment position. Although the Enlightenment thinkers rejected the vision of a providential order, they remained committed to the life goods supported by the Deist order: the ideal of self-responsible reason, the central significance of ordinary fulfillments (production and family life), and the ideal of universal and impartial benevolence.[13] The Enlightenment thinkers not only espoused these goods, they saw themselves as their most effective defenders. Radical disengagement was seen as enhancing self-responsible reason. The instrumentalist stance promised more effective pursuit of happiness and more effective promotion of universal benevolence. Taylor summarizes the new moral view and its power:

13. "Life goods" is a term of art employed by Taylor to signify those goods that provide the framework within which we make individual judgments about what is good or valuable. I will look at this dimension of his argument in more detail in the next section of this chapter.

In giving central significance to sensual pleasure and pain, and in challenging the different conceptions of order, the utilitarians made it possible for the first time to put the relief of suffering, human but also animal, at the center of the social agenda. This has had truly revolutionary effects in modern society, transforming not only our legal system but the whole range of our practices and concerns.[14]

According to Taylor, therefore, an accurate picture of the Enlightenment must account for both its epistemological and moral motivations.

It is here that Taylor's assessment of the Enlightenment worldview is most pertinent when coming to grips with the ecological crisis. He sees a major problem in the Enlightenment position, arguing that it is shot through with contradiction. Contained in its rejection of providential order, its adoption of new understandings of reason and freedom, and its assumption that the relief of suffering *matters* is an impassioned commitment to three central life goods: the ideal of self-responsible reason, the pursuit of ordinary happiness, and the ideal of universal benevolence. Yet these moral motivations cannot be stated within the terms of its theory of human nature. The strong Enlightenment position reduces all human motivations to pleasure or pain and leaves no space for what Taylor calls strong evaluation, which he defines as "the recognition that certain goals or ends make a claim on us, are incommensurable with our other desires and purposes."[15] The language of pleasure and pain will not adequately express the depth of the ideals of self-responsible reason and universal benevolence. Because these life goods cannot be stated within its theory of human nature, they become hidden from sight in contemporary self-understanding. And with these life goods hidden from sight, the instrumentalist worldview has the possibility of destroying the very goals that originally motivated it.

Some who hold the Enlightenment view think that they resolve the problem of its inarticulacy about the good by showing an ultimate harmony of interests between an individual's pleasure and that of all humanity, or showing that humans are moved by sympathy for each other's well-being.[16] This is a naïve argument in itself, but more importantly, it fails to answer the question: Why *ought* I seek universal benevolence? Its understanding of the human person remains unable to articulate its ethic.

In summary, the instrumentalist understanding of the natural and the human world, inherited from the Enlightenment, is unable to account for the very life goods that motivate it. In the company of other significant twentieth-century philosophers, such as Heidegger, Gadamer, and Habermas,

14. Ibid., 331.
15. Ibid., 332.
16. Ibid., 336.

one of Taylor's basic theses is that humans are "self-interpreting animals."[17] By this he does not simply mean that we have a compulsion to be reflective, but rather that we are partly constituted by our self-interpretation. Our understanding of ourselves and our world shapes both our lives and the world around us. When considering the understandings that govern Western culture, therefore, the fact that the goods that motivate the instrumentalist worldview cannot be articulated is much more than an unfortunate occurrence. If they cannot be articulated, they lose their power to shape the world. Further, not only are local goods occluded in this worldview, there is no possibility of identifying a constitutive good, the love of which enables us to do and be good. But that point will be developed more fully in the section that follows.

Alongside the Enlightenment there is another major influence on contemporary self-understanding. Taylor identifies this influence as late eighteenth-century Romanticism.[18] Romanticism emerged in reaction to the instrumentalism of the Enlightenment, yet even though these two views conflict, Taylor sees them as related in a complex way in contemporary culture. For Taylor, the central philosophical insight of Romanticism is the understanding of nature as a source. He says, "This notion of an inner voice or impulse, the idea that we find the truth within us, and in particular in our feelings—these were the crucially justifying concepts of the romantic rebellion in its various forms."[19] A key factor that distinguishes this understanding of the relationship between humanity and nature from that of the Deists is that the voice of nature is now within. Human freedom is understood in terms of the person fulfilling the impulse of nature within.

In Taylor's view, then, the situation of contemporary Westerners is summarized by the image of living in three-dimensional space. He says, "Modern moral culture . . . can be schematized as a space in which one can move in three directions. There are two independent frontiers and the original theistic foundation."[20] These two independent frontiers are first, the Enlightenment worldview, which prizes the dignity of disengaged reason and its associated instrumental view of nature, and second, the worldview of Romanticism, in which the goodness of nature and the power of expressive articulation are strongly valued. Today we are still living out the implications of the Enlightenment and Romanticism or exploring their possibilities.

17. See Taylor, "Self-Interpreting Animals," in *Human Agency and Language*, Philosophical Papers 1 (Cambridge, U.K.: Cambridge University Press, 1985) 45–76.

18. See *Sources of the Self*, chaps. 20–21, pp. 355–390.

19. Ibid., 368–369.

20. Ibid., 317. See also *Sources of the Self*, 496–497.

Overcoming Instrumentalism

If they accept Taylor's view of the modern identity, those who want to protect the earth from ecological devastation may wish to exorcise the Enlightenment worldview from contemporary consciousness and opt for a solely Romantic understanding of nature or one of its successors. But such a summary dismissal is neither possible nor desirable. Taylor argues that the Enlightenment worldview is so deeply ingrained in our self-understanding that we are unable to purge it from our lives. Neither would such action be desirable. He points out that some of the great advances in Western society are due to this strand of the modern identity. As inheritors of the Enlightenment worldview, we feel particularly strongly the imperative to reduce suffering, the importance of universal justice, the centrality of the self-determining subject and of the equality of persons.[21] Simple solutions like the attempt to dismiss the Enlightenment strand from the modern identity cannot account for the complexity of the modern identity, and those who propose these solutions usually still rely on some version of what they deny. Taylor puts it in this way:

> Those who condemn the fruits of disengaged reason in technological society or political atomism make the world simpler than it is when they see their opponents as motivated by a drive to "dominate nature" or to deny all dependence on others, and in fact conveniently occlude the complex connections in the modern understanding of the self between disengagement and self-responsible reason and the affirmation of ordinary life. Those who flaunt the most radical denials and repudiations of selective facets of the modern identity generally go on living by variants of what they deny.[22]

In Taylor's view, the instrumentalist stance is deeply confused. One of its strongest moral motivations, although unarticulated, is the desire to reduce suffering, yet its articulated views of human agency and nature place the planet in grave danger. And because the moral motivation is read out of the official story, this moral source loses its power to motivate—that is, when the desire to reduce suffering is not kept in front of our minds, fulfilling that desire does not retain its significance for us. Taylor's point is that to adequately describe themselves, instrumentalists rely on the moral impetus to reduce suffering, yet they are unable to articulate this moral impetus, and their actions destroy the very thing that they assume. What is required to overcome instrumentalism, in Taylor's view, is a work of retrieval: "to uncover buried goods through rearticulation—and thereby

21. See ibid., 394–397.
22. Ibid., 504.

to make these sources again empower, to bring the air back again in the half-collapsed lungs of the spirit."[23]

Articulacy about the good matters because it is a necessary condition of adhering to what is good; but this connection between life and the good has been lost in much modern moral philosophy. Taylor's philosophical argument about the nature of human agency runs alongside his history of the modern identity. He develops the philosophical argument in Part 1 of *Sources of the Self* and in the process addresses alternative understandings of identity and the moral life. In the following paragraphs I will outline Taylor's understanding of identity around the key concepts of frameworks, defining community and constitutive goods.

In the process of coming to a sense of identity, we define ourselves in terms of what is of crucial importance to us or what we strongly value. That is, in coming to understand ourselves as humans, we necessarily regard some things as significant or worthwhile and disregard others. Taylor calls these strong values "life goods." It is possible to consider these life goods simply as things to which we are strongly attracted or attached, but that would be to miss a crucial dimension of them: they provide the means by which we are able to evaluate experiences, to determine what is good, valuable, worthwhile, or admirable in the events of life. Our identity, therefore, is defined by strongly valued goods, and these goods provide the *framework* or horizon within which we make individual judgments about what is good or valuable.[24] To know who we are is to be situated within a framework of meaning. We are necessarily oriented in moral space.

Although judgments about identity are made by individuals, they are never achieved in isolation. We come to understand ourselves in relation to those who surround us. Here Taylor's understanding of the human as a language being is significant.[25] Without language we would be less than human. We achieve personhood through being initiated into a language. Yet a language only exists and is maintained within a language community. In terms of the question about identity, therefore, I am a self among other selves and cannot be a self alone. We are selves in dialogue. In Taylor's words, "The full definition of somebody's identity thus usually involves not only his stand on moral and spiritual matters, but also some reference to a *defining community*."[26]

---

23. Ibid., 520.
24. Ibid., 51.
25. Taylor, "Language and Human Nature," in *Human Agency and Language*, 217. See also "Theories of Meaning" in the same volume; and *Sources of the Self*, 32–40.
26. *Sources of the Self*, 36. Emphasis added.

Life goods give meaning to human existence and provide a framework within which to make individual judgments. "Good" here designates whatever is considered admirable or worthy. But Taylor argues that there is a good in a fuller sense which gives meaning to or constitutes the goodness of life goods. Life goods direct us toward some feature of the way things are, in virtue of which these life goods are good. He calls this feature a "constitutive good." In his words, "The constitutive good is a moral source . . . it is something the love of which empowers us to do and be good. . . . The constitutive good does more than just define the content of the moral theory. Love of it is what empowers us to be good. And hence also loving it is part of what it is to be a good human being."[27]

Taylor identifies three constitutive goods at work in the modern Western world: (1) the Idea of the Good itself is the key to Plato's order; (2) the Christian notion of *agape* as a love that God has for humans, in which they participate through grace; (3) even though contemporary Kantians would balk at the idea of a constitutive good, Taylor argues that just as their position is grounded on an unadmitted adherence to the life goods of freedom and justice, so too rational agency acts as a constitutive good in their worldview.[28]

In summary, according to Taylor, every person lives in the framework of life goods, which are identified within a defining community and refer to a constitutive good, the love of which empowers them to do and be good. This philosophical understanding of personhood goes hand in hand with his historical picture of the modern identity and his analysis of instrumentalism.

Since late eighteenth-century Romanticism there has been a range of responses to the instrumental mode of life. Taylor argues that instrumentalism has been criticized on two primary levels: experiential and public. First, at the experiential level, instrumentalism has been criticized by a wide range of thinkers, including Tocqueville, Nietzsche, Weber, and Arendt. They argue that the instrumental mode of life empties life of meaning. It leaves no room for heroism or a powerful sense of purpose, it results in disenchantment and fragmentation, or it destroys the matrices of meaning.[29] Second, the instrumental mode of life is criticized for its strongly negative public consequences. From an ecological perspective, the instrumentalist mode of life places the long-term future of the planet at risk. It also threatens public freedom.[30] When self-rule depends upon a

27. Ibid., 93.
28. Ibid.
29. For a fuller description of these perspectives see *Sources of the Self,* 500–502.
30. Ibid., 502.

common sense of purpose, this is radically undermined by a worldview that cannot admit to the existence of common or shared goods.

In reference to these two levels of critique, Taylor examines a range of analyses of the problems of instrumentalism.[31] First, he considers the position of those who stoutly resist the critique and continue to understand the subject as a disengaged entity standing over nature. In dismissing the criticism, these protagonists argue that the experiential consequences of instrumentalism are illusory and that the public consequences can only be resolved with a further dose of instrumentalist reason. Second, he considers the position of the modern-day heirs of Romanticism, like the human potential movement. Third, Taylor addresses Jürgen Habermas's reading of instrumentalism and the modern identity several times in *Sources of the Self* and in other places.[32] Central to his response to each of these readings of instrumentalism is the claim that they all fail to recognize or too readily dismiss some goods that inform the modern identity—goods by which we live. For instance, Taylor points out, the disengaged rationalists, even though they deny the experiential consequences of instrumentalism, still find themselves struggling with the dilemmas of their personal life, using concepts like fulfillment.

Taylor's primary line of argument in *Sources of the Self* is that modernity is rich in life goods or moral sources. He argues, therefore, that we can only adequately account for modern self-understanding in terms of the moral sources of the Enlightenment and Romanticism: self-responsible reason, the pursuit of ordinary happiness, the ideal of universal benevolence, and the ideal of nature as a source. Yet this moral richness is often rendered invisible by the instrumentalist mode of thought. And this is the intuition that inspired *Sources of the Self:* that "we tend in our culture to stifle the spirit."[33] As a result, some of the key problems of our age can be solved only through uncovering by rearticulation some of the buried goods on which we rely to make sense of ourselves and allow these sources to inspire us.

In Taylor's view, therefore, there are two crucial dimensions to overcoming instrumentalism and finding a solution to the ecological crisis. Because our self-understanding has a profound impact on the way in which we interact with and shape the world around us, what is required is a work of retrieval. This work of retrieval would first articulate the claims that nature

31. Ibid., 504–511.
32. See *Sources of the Self,* 85–88, 509–510. See also his essay "Language and Society," in *Communicative Action: Essays on Jürgen Habermas's The Theory of Communicative Action,* ed. Axel Honneth and Hans Joas, trans. Jeremy Gaines and Doris L. Jones (Cambridge, U.K.: Polity Press, 1991) 23–35. Habermas responds to Taylor in the final chapter of the volume.
33. *Sources of the Self,* 520.

makes on us. Taylor avers that "it would greatly help in staving off ecological disaster if we could recover a sense of the demand that our natural surroundings and wilderness make on us."[34] Second, the work of retrieval would articulate an understanding of human agency that accounts for humanity's enmeshment with life goods and a constitutive good. In the remainder of this chapter I want to consider these points. I want to show that Karl Rahner's understanding of the human person in relationship to the human community, the earth, and to God supports the positive ideals of the Enlightenment—self-responsible reason and freedom. I hope to show, therefore, that the instrumentalist view of agency is not only wrongheaded, but it is not necessary as a support to self-responsible reason and freedom.[35]

Karl Rahner: Humanity in the Presence of Mystery

A constant theme of Karl Rahner's theology is that humanity does not stumble upon God by accident. Rahner argues that humans, by nature, are oriented toward absolute mystery. All the time and in everyday ways we reach out to mystery. In every act of knowing what is true or loving what is good, there is an implicit or unthematic relationship to mystery. His early philosophical works *Spirit in the World* and *Hearer of the Word* spell out this relationship through an examination of the nature of human knowledge and freedom.[36] In later articles on the experience of the Spirit, he develops the relationship between human experience and the experience of God.[37] Toward the end of his life Rahner explored the human orientation to mystery in the context of the quest for meaning.[38] In the following paragraphs I will outline the connection he makes between meaning and mystery.

One common way of expressing the contemporary search to make sense of our lives (what I have been calling the search for self-understanding) is in terms of the *meaning* of life. We speak of meaning in many ways. We seek to grasp the meaning of particular experiences or events, for example, a

34. Ibid., 513. Taylor uses Martin Heidegger's philosophy of language to explore the claims that the natural world makes on us in "Heidegger, Language and Ecology," *Philosophical Arguments* (Cambridge, Mass.: Harvard University Press, 1995) 100–126.

35. See *Sources of the Self*, 514.

36. Rahner, *Spirit in the World*, trans. William V. Dych (London: Sheed & Ward, 1968); Rahner, *Hearer of the Word: Laying the Foundation for a Philosophy of Religion*, trans. Joseph Donceel (New York: Continuum, 1994).

37. See Rahner, *Theological Investigations* 16: *Experience of the Spirit: Source of Theology*, trans. David Morland, O.S.B. (New York: Crossroad, 1983).

38. See Rahner, "The Question of Meaning as a Question of God," in *Theological Investigations* 21, and "The Human Question of Meaning in the Face of Absolute Mystery," in *Theological Investigations* 21: *God and Revelation*, trans. Edward Quinn (London: Darton, Longman & Todd, 1984) 89–104.

political event, a book, a life, or a relationship. When discussing meaning here, however, I am not referring to partial experiences of meaning; rather, I want to reflect on the definitive meaning of human existence itself. I want to explore Karl Rahner's answer to the question: What is it that gives human existence meaning overall? In summary, Rahner argues that humans grasp the ultimate meaning of existence when they find themselves in the presence of absolute mystery. Clarity about two key points is essential to understanding Rahner's position: his notion of mystery and his understanding of the human orientation to mystery.

Rahner criticizes the notion of mystery common in the neo-Scholastic theology of the nineteenth and early twentieth centuries.[39] He regards the prevalent neo-Scholastic understanding of mystery as not necessarily in error but as deficient. In the theology of the schools, mysteries were ordinarily regarded as truths (i.e., statements) that are impenetrable for the moment but finally, perhaps in eternity, would succumb to human reason. Rahner notes that the unacknowledged presupposition of this view of mystery is that "we are dealing with truths which should strictly speaking have come within the scope of reason with its power to see and *comprehend*, but in this case do not meet its demands."[40] He argues that this view of mystery is not faithful to the wider theological tradition.

For Rahner, mystery must be understood in the context of the incomprehensibility of God. Referring to sources as diverse as Leo the Great and Vatican I, Rahner points out that in the theological tradition God is understood as incomprehensible by nature.[41] Because God is the infinite source of our being, God is more than our finite, created intellect can bear. And when God communicates Godself, this communication necessarily shares in the incomprehensibility that belongs to God, since God communicates Godself and not something else. Rahner succinctly expresses the link between incomprehensibility and mystery in this way: "The incomprehensible God comes to us as mystery."[42] Mystery, then, is the depth and richness of God that continually opens up to us rather than an answer that temporarily eludes us.

Central to Rahner's understanding of the human orientation to mystery is a strong critique (similar to Taylor's) of the Enlightenment view of knowledge. Rahner criticizes this view because in it knowledge is regarded as the

---

39. Rahner, "The Concept of Mystery in Catholic Theology," in *Theological Investigations* 4: *More Recent Writings*, trans. Kevin Smyth (London: Darton, Longman & Todd, 1966) 37–39.

40. Ibid., 39.

41. See Rahner, "The Human Question of Meaning," 90–92.

42. Rahner, "Mystery," in *Sacramentum Mundi: An Encyclopedia of Theology* (London: Burns & Oates, 1969) 4:134.

capacity to see through reality in order to dominate it.[43] He argues that the dynamism of human knowing consists in the intellect's grasping a concrete object while reaching beyond to being in its totality.[44] In the act of knowing, the person is always at once in the world and also reaching out to absolute being. In Rahner's words, when reason "grasps and understands any object, it has already transcended the latter into an infinity that is present as unexplored . . . it always seizes the individual object by being tacitly aware of the fact that the object always is and remains more than what is grasped of it."[45] So, according to Rahner, the intellectual tasks of elaborating, delimiting, and discriminating—those tasks on which the Enlightenment view focuses—occur while the human reaches out to the incomprehensible.

It is Rahner's argument, therefore, that because of the dynamism of the knowing subject, humanity is oriented to absolute mystery. Reason, he says, is "the power of coming face to face with incomprehensibility itself."[46] This is more than a claim that such a concept of reason is appropriate for a theological view of the world—it is a philosophical claim about the nature of reason itself. "Reason is of this nature," Rahner says, as he argues for its relationship to mystery and dismantles a rationalistic theory of knowledge.[47] He argues that a similar dynamic of the human reaching out to mystery occurs in human freedom.[48]

In further considering the nature of knowledge and freedom, Rahner concludes that when we recognize our orientation to mystery and accept the incomprehensible God in self-surrendering love, it is then that we grasp the ultimate meaning of life.

From a purely philosophical perspective, one that prescinds from the data of Christian revelation, the mystery toward which our hearts and minds reach could be understood as distant and aloof, never unambiguously revealing itself. Outside the context of revelation, human transcendence to mystery leaves a question about the nature of that to which it reaches. Need mystery always remain distant? Here Rahner turns to the Christian doctrines of grace and incarnation and shows that they reveal the nature of holy mystery. In grace—God's self-communication—the mystery comes close. Grace is primarily the self-giving of the mystery. Yet in this self-giving the mystery remains essentially mysterious and incomprehensible. Rahner puts

43. Rahner "The Human Question of Meaning," 95.
44. Rahner explores this view of knowledge in *Spirit in the World*.
45. Rahner "The Human Question of Meaning," 98.
46. Ibid., 99.
47. Ibid., 97.
48. I have traced the lines of his concept of freedom in *Freedom in the World: The Significance of Karl Rahner's Theology of Freedom in the Light of Charles Taylor's View of the Modern Identity*, unpublished Ph.D. thesis, Flinders University, 1994.

it this way: "Grace is . . . the grace of the *nearness* of the *abiding* mystery: it makes God accessible in the form of the holy mystery and presents him thus as the incomprehensible."[49] Rahner coins the term "supernatural existential" to express the Christian belief that every human person lives within the offer of God's grace. God's self-communication is never merited and therefore, in that sense, is not natural to humanity. It is absolutely gratuitous, but nonetheless the fulfillment of human knowledge and freedom.

The self-communication of the mystery takes on its most radical form in the incarnation. In the Word made flesh, the self-transcendence of the human spirit is fulfilled by the self-gift of absolute mystery. The mystery of the incarnation, therefore, is not a mystery in any other sense than the absolute closeness of God as holy mystery—"it is this mystery itself in an unsurpassable form."[50]

Rahner's theology of the human person in the presence of mystery, as I have described it so far, could be regarded as anthropocentric. He certainly adopts modern philosophy's anthropological turn. Anne Carr describes his theology as beginning with the question: "What light can be shed on the question of the human person by an analysis of what the Christian message presumes as true about that person?"[51] Yet, simplistically equating the anthropological turn with an anthropocentric theology or philosophy would result in a misreading of both Rahner's theology and much contemporary philosophy.[52] Clarity about humanity's relationship to the rest of creation is essential here. Rahner holds that humanity has a significant and unique place in the cosmos, because through humanity the cosmos comes to consciousness.[53] However, although holding a unique place, humanity remains profoundly linked to the rest of creation. In Rahner's view, the drive toward consciousness is of the very nature of the universe itself. Every created thing exists for the self-communication of God. He says, "Every natural created entity is ordered to this grace in such a way that it cannot remain really whole and healthy in itself, nor achieve the completion required by its own nature, except as integrated into the supernatural order of grace."[54]

49. Rahner, "The Concept of Mystery," 56. Emphasis in original.

50. Ibid., 69.

51. Anne Carr, "Starting with the Human," in *A World of Grace: An Introduction to the Themes and Foundations of Karl Rahner's Theology,* ed. Leo J. O'Donovan (New York: Seabury Press, 1980) 17–18.

52. Taylor, for example, acknowledges the significance of the anthropological turn in modern thought and yet argues against an anthropocentric view of the cosmos. See "Heidegger, Language and Ecology."

53. See Rahner, *Foundations of Christian Faith,* 188–189.

54. Rahner, "The Order of Creation and the Order of Redemption," in *A Rahner Reader,* ed. Gerald A. McCool (London: Darton, Longman & Todd, 1975) 194.

Living in the presence of mystery requires us to be open to the presence of mystery in all things. With regard to other humans, this openness is usually articulated in terms of the theological virtue of love.[55] In the love of another we open ourselves to the presence of mystery. The recognition that all creation is ordered to mystery calls us to love all of creation. The natural world makes demands on us because it is more than merely natural, more than something to be made over—it reveals the mystery of God.

In the second section of this chapter, I outlined Charles Taylor's line of thought, leading to his judgment that crucial to overcoming the ecological crisis is the articulation of a view of human agency that accounts for humanity's enmeshment with life goods and a constitutive good. It is my argument that Rahner's theology of the human person in the presence of absolute mystery does precisely this. From an understanding of knowledge and freedom that values the "turn to the subject," Rahner argues that God, the Mystery, is the constitutive source of our lives, to use Taylor's words. And because we humans are material and social beings, in Rahner's view we encounter God in the ordinary experiences of life—in acts of knowledge and freedom. We encounter the mystery in one another and in our relationship to Earth.

Planet Earth is in grave danger not simply because humans have invented tools of destruction. Earth is at risk because of the way humanity understands its relationship with Earth. Our precarious state is the result of the influential Enlightenment views both of nature as neutral (existing to be made over by humans) and of the human person as disengaged, standing over nature in order to master it. An adequate solution to the crisis requires us to deal with the root of the problem. Solutions that look primarily to further scientific advances for remedy will not address the problem at its root. When, as humans, we recognize that we are not disengaged from, but rather profoundly connected to, all forms of life on the planet, and when we recognize that only through our relationship with the rest of creation can we enter into relationship with the divine Mystery, then we are well on the way to living in a new relationship with creation and healing planet Earth. Such a renewed relational understanding will reveal a deeper, truer understanding of human existence and particularly of reason and freedom as situated within creation and the divine Mystery.

---

55. See Rahner's seminal essay "Reflections on the Unity of the Love of Neighbour and the Love of God," in *Theological Investigations* 6: *Concerning Vatican Council II,* trans. Karl-H. and Boniface Kruger (London: Darton, Longman & Todd, 1969) 211–249.

Dancing Man

He's dancing.
He's happy
Because he's caught a big feed.
Now him and his family
Are gunna have a big feast.

Tami Munnich

Chapter 12

Bioethics, Ecology, and Theology

Andrew Dutney

The word "bioethics" immediately brings to mind the medico-scientific sphere. We think first of research, experimentation, and the use of technology in health care, of things like genetic therapies, the latest wonder drug, cloning techniques, or the ethical monitoring of such portentous advances in medicine. After that we would probably remember that, around this central focus, bioethics also relates to aspects of clinical practice, where it informs what might be better understood as professional ethics for health care workers. It offers guidance to doctors, nurses, and others who are attempting to provide services that are fair, respectful, and beneficial for each patient. On further thought, we might even grant that a more distant relation of bioethics can be found in the sphere of community health. We might, for example, think of the ethical dimensions of the correlation between health and ethnicity, gender, age, or economic status. However, here the medico-scientific focus of real bioethics might seem to be compromised by the preoccupation with social and political ethics.

In other words, the common understanding of bioethics places medical science at the center, orients the care of individuals around that center, cautiously admits limited consideration of human communities, and ignores entirely the natural environment. That is what bioethics usually means.[1]

---

1. So, for example, one major text asserts that "in the middle of the sixties public attention was drawn to a series of hard moral questions about medicine . . . and bioethics was born." Allen Verhey and Stephen E. Lammers, "Rediscovering Religious Traditions in Medical Ethics," in *Theological Voices in Medical Ethics,* ed. Allen Verhey and Stephen E. Lammers (Grand Rapids, Mich.: Wm. B. Eerdmans, 1993) 1–6, at 1. The uncontentiousness of

This essay questions that common understanding of bioethics, which is historically inaccurate, conceptually flawed, and has been largely over-taken by events. Moreover, from the perspective of our Ecotheology Pro-ject, it is evident that present global circumstances invite from the Church a different kind of bioethics.

A New Medical Ethics

During the 1960s a series of momentous developments in medical science and technology was provoking a thorough revision of the tradi-tional field of "medical ethics." The general public was presented with a series of "medical miracles," such as kidney dialysis, organ transplanta-tion, artificial life support, the contraceptive pill, prenatal diagnosis, and medically safe abortion. But the beginning of the medical revolution brought with it profound challenges to community life and values. How do we prioritize patient access to dialysis or potential organ recipients? What is an ethically legitimate process for recruiting organ donors? When is a person "dead," so that their organs may be harvested? How long should we keep a person on life support? What is a "person" anyway? Can we agree on values that will help people make their reproductive choices? Do we have any choice about having all these choices put before us? To whom should medical researchers be accountable for the methods by which they achieve these advances? What kind of legislation is needed to protect the vulnerable in the new areas that are opening up?

Medical ethics had a long tradition, but it had been suffering from general neglect for most of the century and, in any case, had been so nar-rowly focused on the professional responsibilities of doctors that it could not offer the flexibility and subtlety of analysis demanded in the new con-text. Some other approach was needed.

Even though it was published before there was any real awareness of the startling changes that were coming upon medicine, Joseph Fletcher's book *Morals and Medicine* (1955) is widely recognized as marking the turning point in medical ethics.[2] Fletcher, an Episcopalian theologian, de-

---

this definition of "bioethics" is confirmed by scanning the contents pages of standard text-books, such as *Bioethics: Basic Writings on the Key Ethical Questions That Surround the Major, Modern Biological Possibilities and Problems,* 4th ed., ed. Thomas A. Shannon (Mahwah, N.J.: Paulist Press, 1993) v–viii.

2. Daniel Callahan, "Religion and the Secularization of Bioethics," in *Hastings Cen-ter Report* 20 (July–August 1991), Special Supplement on Theology, Religious Traditions and Bioethics, ed. Daniel Callahan and Courtney S. Campbell (Hastings-on-Hudson, N.Y.: The Hastings Center, 1991) 2–4, at 3.

cried the separation of human values from scientific facts, the isolation of morality from medicine. If nothing else, it had led to the stagnation of the field of medical ethics:

> Medicine and human beings have suffered enough in the past because there were religionists who liked to say, and doctors willing to let them say, that morals in medicine are "a matter for the church, not for doctors." . . . Such cynicism accounts, perhaps, for the significant fact that the catalogues of medical libraries in this land turn up a great many titles on medical ethics written in the nineteenth century, but relatively few for the twentieth.[3]

Fletcher challenged the shift to a kind of "value free" medicine that he considered to have been taking place during the twentieth century. He argued that the central problem in medical practice was the failure to distinguish morally between persons and their bodies. Medicine had tended toward the treatment of bodies and had relied by default on "physical, or physiological, standards of morality," including versions of natural law in particular. It was Fletcher's contention that a genuinely ethical focus for clinical practice required that priority be given to "persons."

The final chapter of *Morals and Medicine* was devoted to an explanation of Fletcher's "personalist" ethic. In particular, it concerned the interpretation of the relation between nature and human nature. Fletcher argued for the recognition of an essential dualism within the created order, distinguishing humanity from nature, and the person from the body. Completing his hierarchy of being, Fletcher was careful to distinguish God from humanity, affirming that "men [sic], as men, are as yet only neophytes in the spiritual world and the forum of conscience."[4] The fundamental issue for Fletcher was that "physical nature is what is over against us, out there."[5] Submission to nature was seen as a failure of moral responsibility, a retreat from the call to be human, and an ethic oriented to natural law was no more than a systematic restraint upon spiritual life and growth, "a perversion of a moral norm into a physical or material norm."[6] So emphatic was the argument offered in that chapter that it could be said justifiably that Fletcher's book "celebrated the power of modern medicine to liberate human beings from the iron grip of nature."[7]

3. Joseph Fletcher, *Morals and Medicine* (London: Victor Gollancz Limited, 1955) xiii. I note that the medical library in our university has a very modest collection on ethics, certainly much smaller than the same section in the theological library.

4. Ibid., 219.

5. Ibid., 211.

6. Ibid., 222.

7. Daniel Callahan, "Bioethics," in *Encyclopedia of Bioethics,* rev. ed., ed. Warren Thomas Reich (New York: Simon & Schuster, 1995) 247–256, at 249.

Yet this obscures an important qualification that Fletcher inserted early in the discussion. He wrote, "It would be wrong to give the impression that nature, in our own human physique or in any other part of the natural order, is merely indifferent material to be used or discarded without respect to its constitution, its limits, order, and trend. We must repeat that man is himself a creature of the natural order."[8] Fletcher applied the principle of human "partnership" with nature to the relation of the person to the body. The principle had been articulated by William Temple, the great Anglican theologian and archbishop of Canterbury.[9] For Fletcher, the body "is not a mere servant, to be manipulated and ordered about as having no claims or worth of its own," but neither is it "an overlord." So he offered two principles: "(1) we ought not to submit willy-nilly to what is, to physical and physiological facts simply as they are, since to do so would be to be unfree; and (2) we ought not to ignore or disregard or flout what is, simply because it is unchosen, since to do so would be to be guilty of an unrealistic denial of human fini-tude and creaturehood, and therefore irresponsible."[10]

It must still be granted, however, that these principles were not weighted equally by Fletcher. The good of "freedom" requires that it al-ways be maximized, while the legitimate restraint of "what is" should be minimized and, indeed, further reduced with human "progress," especially progress in medical science and technology.

Fifteen years later Paul Ramsey, another theologian, published the first systematic treatment of medical ethics from the midst of the new age of medical miracles.[11] The title of his book, *The Patient as Person,* seemed to reflect an acceptance of Fletcher's central argument. However, with the medico-scientific revolution now well underway, Ramsey found it necessary to place emphasis on the other side of personhood. Fletcher had leaned in

8. Fletcher, *Morals and Medicine,* 212.

9. William Temple, *The Hope of a New World* (London: SCM Press, 1940) 67: "Man is a part of the system of nature, whatever else he may be beside. He must study the ways of nature and follow them, for he is utterly dependent on the natural world. Consequently, he must not think of natural resources as there for him to exploit to his own immediate ad-vantage, but must rather co-operate with the natural process and so, in the long run, gain a far greater advantage. This is of primary importance in relation to man's treatment of the soil. Nature is man's partner rather than his servant; he is dependent on it for the means of life. For the Christian this is recognized as a pact of creatureship. The treatment of the earth by man the exploiter is not only imprudent but sacrilegious. We are not likely to correct our hideous mistakes in this realm unless we can recover the mystical sense of our one-ness with nature. I labour this precisely because many people think it fantastic; I think it is fun-damental to sanity."

10. Fletcher, *Morals and Medicine,* 213.

11. Paul Ramsey, *The Patient as Person* (New Haven and London: Yale University Press, 1970).

the direction of personal freedom, but Ramsey, now surrounded by the new medicine, shifted the weight toward "human finitude and creaturehood." Ramsey's view of personhood had two main foci. One was the "biblical" view of the person as essentially *embodied,* "a very realistic view of the life of man [sic] who is altogether flesh *(sarx)*."[12] In terms that could be taken as a mild reprimand to Fletcher, or at least to Fletcher's less careful admirers, he wrote,

> From this point of view, one must ask of any Christian, who today without any hesitation flies into the wild blue yonder of transcendent human spiritual achievement while submitting the body unlimitedly to medical and other technologies, whether his outlook is not rather a product of Cartesian mentalism and dualism, and one that, for all its religious and personalistic terminology, has no longer any Biblical comprehension of joy in creaturely life and the acceptable death of all who are flesh.[13]

This final phrase, "joy in creaturely life and the acceptable death of all who are flesh," summarized in Christian terms the good toward which medical ethics was to be directed. It was a good that owned, and even celebrated, the necessarily embodied nature of personhood. The other focus of Ramsey's view of personhood was its essential *relationality*. In the preface to the book, he proposed a covenantal model of ethics based on the covenantal nature of the human person. He wrote:

> We are born within covenants of life with life. By nature, choice, or need we live with our fellowmen in roles or relations. Therefore we must ask, What is the meaning of the faithfulness of one human being to another in every one of these relations? This is the ethical question.[14]

So by the early seventies a new medical ethics was appearing with several key themes: a rejection of the possibility of "value free" science, a questioning of triumphalist humanism and its ideology of progress, and a view of the human person as embodied and relational.

In the case of the two pioneers Fletcher and Ramsey, these themes were developed in ways that included explicit references to their theological underpinnings. Fletcher drew directly on William Temple's principle of human "partnership" with nature, a minor theme in Protestant theology that

12. Ibid., 187. It must be said that Ramsey's view of personhood is not properly unpacked in that book but must be inferred from asides made during his more sustained discussions of particular issues in medical ethics.

13. Ibid., 188

14. Ibid., xii. Yet even as Ramsey explained this "biblical norm" and its relevance to medical ethics, he admitted that it "is not a very prominent feature in the pages that follow." He was as good as his word!

appears consistently from the nineteenth century, usually in the context of a critique of the despoliation of nature by industry and an appeal to Christian capitalists for moderation.[15] For Fletcher it was a formal exercise, a conventional theological gesture in the direction of restraint on *hubris* in medical progress. Ramsey's theological sources were not explicit but obvious nonetheless. His use of biblical theology and of covenantal themes in particular located him in the Protestant neo-orthodox movement, which was at the height of its influence in the middle of the twentieth century. There was something essentially formal about Ramsey's use of this theological resource too. While Fletcher had used conventional liberal theology in the cause of *restraining* excesses in medical progress, Ramsey used neo-orthodox motifs in the cause of *correcting* medical and research practices. Both drew on existing theological sources and used them in the conventional ways. That is, to the limited extent that they made theological references in their projects, they received their theological resources uncritically, as givens, and treated medical ethics as *applied* theology.

However, the connection to the Christian tradition was entirely secondary. The purpose of the new medical ethics was to ensure that the ongoing work of medical research and technological innovation would be guided and moderated by human values, and by a constant awareness of the personhood of those routinely identified as "subjects" and "patients." The Church's only stake in this endeavor was to be of service to the community in a time of great change. It had no interest in advertising its association with the development of the new medical ethics.

Bioethics

It took some time for the new medical ethics to be called "bioethics." The first use of the term in this sense was in 1971, when the Kennedy Institute of Human Reproduction and Bioethics was founded at Georgetown

---

15. See, for example, Newman Smyth, *Christian Ethics,* 3rd ed. (Edinburgh: T&T Clark, 1892) 320–324, and Brooke Foss Westcott, *Christian Aspects of Life* (London and New York: Macmillan, 1897) 369–383, at 374–376. Albert Schweitzer's famous ethic of "reverence for life" belongs to this tradition too, albeit in an intensified and more profound form. See, for example, his *Civilization and Ethics,* 3rd ed. (London: Adam & Charles Black, 1949 [1923]) 240–280, at 252–254. And he was not alone in recognizing that a critique of the worldview that had Europe stumbling horribly into the Great War would include a theological reconsideration of the relationship between humanity and nature. See, for example, E. Griffith-Jones, *The Dominion of Man: Some Problems in Human Providence* (London: Hodder and Stoughton, 1926) 226–238, and H. H. Farmer, *The World and God: A Study on Prayer, Providence and Miracle in Christian Experience* (London: Nisbet & Co., 1935) 303–306. This stream of thought provides the context for Temple's principle of partnership with nature.

University, under the leadership of André Hellegers. During the course of that decade, the term gradually caught on, so that the huge project begun in 1971 as the *Encyclopedia of Medical Ethics* was eventually published in 1978 as the *Encyclopedia of Bioethics*. Bioethics was thus confirmed as the successor to "medical ethics."

However, as a matter of fact the term "bioethics" had already been coined and put to use in a quite different area. The word "bioethics" was first used by the oncologist Van Rensselaer Potter in 1970.[16] Although Potter's reputation was based on his pioneering work in medical science, when *Bioethics: Bridge to the Future* was published the following year, it was clear that the new discipline that he proposed had little, if anything, to do with the field of medical ethics. It included chapters entitled "The Role of the Individual in Modern Society," "How Is an Optimum Environment Defined?," "Survival as a Goal for Wisdom," and even "Teilhard de Chardin and the Concept of Purpose." Potter signaled his guiding intention by dedicating the book to the memory of Aldo Leopold, the revered pioneer of environmental ethics and the one who, according to Potter, "anticipated the extension of ethics to bioethics."[17] Then, in the preface Potter explained:

> What we must now face up to is the fact that human ethics cannot be separated from a realistic understanding of ecology in the broadest sense. Ethical values cannot be separated from biological facts. We are in great need of a Land Ethic, a Wildlife Ethic, a Population Ethic, a Consumption Ethic, an Urban Ethic, an International Ethic, a Geriatric Ethic, and so on. All of these problems call for action based on values and biological facts. All of these involve Bioethics, and the survival of the total ecosystem is the test of the value system.[18]

And in case his readers might be put off the scent by the somewhat philosophical approach taken in the body of the book, Potter underscored his intention to identify his project with the new ecopolitical movement by urging that "those who haven't already done so should certainly read *Population,*

16. Van Rensselaer Potter, "Bioethics: The Science of Survival," in *Perspectives in Biology and Medicine* 14 (1970) 127–153. This article was adapted from the first chapter of Potter's book, which was to appear early in the new year: Van Rensselaer Potter, *Bioethics: Bridge to the Future* (Englewood Cliffs, N.J.: Prentice-Hall, 1971).

17. Leopold's *A Sand County Almanac* (London: Oxford University Press, 1949) has been described as "the Holy Writ of American conservationists" by René Dubos in *A God Within* (London: Angus & Robertson, 1972) 156, and as "the environmentalist's bible" by the editor of *Companion to a Sand County Almanac: Interpretive & Critical Essays*, ed. J. Baird Callicott (Madison, Wis.: University of Wisconsin Press, 1987) 3. It has been said that "Leopold would have been both surprised and pleased" at the use of his work in the construction of Potter's "bioethics": Roderick Frazier Nash, *The Rights of Nature: A History of Environmental Ethics* (Madison, Wis.: University of Wisconsin Press, 1989) 85.

18. Potter, *Bioethics*, vii–viii.

Resources, Environment, Issues in Human Ecology by Paul R. Ehrlich and Anne H. Ehrlich and *The Environmental Handbook* by Garret de Bell."[19]

So when the term "bioethics" was coined, its primary reference was to the concerns of environmentalism. The "biological facts" to which "ethical values" were to be connected by this new discipline were not the facts of medicine and health care, but those that were the concern of the life sciences in general—biology, biochemistry, ecology. Bioethics was conceived by Potter as the starting point for developing a "science of survival" to replace the self-defeating science of "growthmanship," which he recognized to have been exposed in Rachel Carson's *Silent Spring*. Trying to carry that debate forward, Potter said, "From many uninformed quarters we now hear demands for a moratorium on science, when what we need is more and better science."[20] It was bioethics that Potter saw as the means to create that "better science" upon which global survival depended.

The term continued to be used in this way well into the eighties. For example, in their influential book *The Liberation of Life*, Charles Birch, a biologist, and John B. Cobb Jr., a theologian, identified their proposals for "an ethic of life" as "bioethics."[21] But, as Potter has acknowledged, this usage "was never recognized by any medical bioethicist to my knowledge, and the word bioethics was narrowly redefined by the medical people to mean clinical ethics. . . . By 1988 it was apparent that the word bioethics was completely out of my hands."[22] So, to return to the example, when John Cobb began his more recent book on bioethics with a chapter on environmental ethics and animal rights and concluded with a chapter on sexual ethics, he needed to concede that this approach now required "a certain

---

19. Ibid., ix–x.

20. Ibid., 3–4 and 42–54. Rachel Carson, *Silent Spring* (Boston: Houghton Mifflin Company, 1962) has a place alongside *A Sand County Almanac* in the memory of the environmental movement.

21. In the index. Charles Birch and John B. Cobb, Jr., *The Liberation of Life: From the Cell to the Community* (Cambridge: Cambridge University Press, 1981) 187, referring to 141–175 and 204–208.

22. Van Rensselaer Potter, "What Does Bioethics Mean?" in *The Ag Bioethics Forum: An Interdisciplinary Newsletter in Agricultural Bioethics* 8 (June 1996) [accessed via http:www.grad-college.iastate.edu/bioethics/forum/forum.html]. This is something of an overstatement, as is illustrated by the title of the journal in which his comment appears. So, too, Potter's project is recognized by Daniel Callahan in his essay on "Bioethics" in the standard reference work, *Encyclopedia of Bioethics*, 250. Indeed, Callahan tries (unsuccessfully) to interpret "bioethics" in a way that may continue to accommodate Potter's project. Warren Thomas Reich has gone further in integrating the two approaches in "The 'Wider View': André Hellegers's Passionate, Integrating Intellect and the Creation of Bioethics," in *Kennedy Institute of Ethics Journal* 9 (1999) 25–51.

stretch of the usual application of the term."[23] For his part, Charles Birch turned to the more transparent "biocentric ethics."[24]

The eclipse of the ecological definition of bioethics is not difficult to understand. Warren Thomas Reich has identified several key factors.[25] The new medical concerns that were called bioethical at Georgetown touched the daily lives of American people in a way that was not true of the more ecological concerns of Potter. At the same time, insofar as questions of civil rights were linked to these new medical concerns—the rights of the handicapped, the poor, women—they became readily accessible to an American public that had been developing a culture of civil rights analysis and debate during the sixties and early seventies. Particular issues, especially those associated with abortion, also generated a political heat that gave Georgetown bioethics a high profile and a certain prestige for its manifest attention to the public good. So, of course, the media developed a habit of seeking expert comment and analysis from this new source as breaking medical stories passed through the editorial process, thus publicizing the Georgetown meaning of bioethics.

Alongside these factors, it was soon realized that the critical analysis of emerging issues in medical practice and research provided engaging material for courses in high schools and colleges, generating a demand for texts in Georgetown's bioethics and graduates with skills in that area. In this context Hellegers was able to marshal uncommitted resources from private, federal, and university sources to operate the new Kennedy Institute and to endow chairs in bioethics. He drew together a community of outstanding scholars from philosophy, the life sciences, and the social sciences to work on a wide range of bioethical projects and to provide consultancies to government agencies and a center for medical research and clinical practice. The graduate program that Hellegers encouraged and supported was soon producing new generations of bioethicists. For his part, Potter did what he could to promote ecological bioethics, but without the support of an institution or academic community dedicated to the project, without the patronage of the mass media, and in addition to his primary work in biochemical research.

Although they concentrated in different areas, medical bioethics and ecological bioethics had some common themes. Both projects rejected the

23. John B. Cobb, Jr., *Matters of Life and Death* (Louisville, Ky.: Westminster/John Knox Press, 1991) 8.

24. In the index. Charles Birch, *On Purpose* (Sydney: University of New South Wales Press, 1990) 187, referring to 131–134.

25. Warren Thomas Reich, "The Word 'Bioethics': The Struggle Over Its Earliest Meanings," in *Kennedy Institute of Ethics Journal* 5 (1995) 19–34, at 21–23.

possibility of "value free" science and represented the determined agitation of those who regarded careful ethical reflection as an essential part of "good science." In this context both were skeptical of the ideology of progress that seemed unable to comprehend the limitations of mortality and sustainability. The view of the human person as embodied and relational, which was emerging in medical bioethics, found a particular expression in ecological bioethics, emphasizing the connectedness of human beings with other species of organisms and, indeed, with the inorganic world.

In the development of the new medical ethics, as pioneered by Fletcher and Ramsey, theological references were explicit but conventional. What was new to the activity was the critical analysis of the reshaped clinical and research context, not the theological elements of that analysis. But the place of theology in the development of bioethics was a little different. In the beginning at least, leading figures in both ecological and medical bioethics tended to espouse varieties of revisionist theology, especially of the sort which resulted from the constructive engagement of theology and science. So, from the ecological perspective, Potter was an enthusiast for Teilhard de Chardin's evolutionary theology, while Birch and Cobb were committed to process theology.

From the medical perspective, Hellegers was deeply involved in new approaches to moral theology emerging from the Second Vatican Council, which were more responsive to scientific and social developments and which were open to more active lay participation. In particular, he was one of the leading figures on the papal birth control commission (1964–1966), which, in spite of the rejection of its recommendations for reform, made an enduring contribution to the period of theological creativity that followed Vatican II.[26] In various ways both ecological bioethics and medical bioethics were treated as something other than *applied* theology. Those involved did more than simply refer to the theological underpinnings of their bioethical projects; they were explicit about the need for theological reform in the face of changing historical circumstances and the growth of scientific knowledge.

Feminist Bioethics

Medical bioethics was intended to be a restraint on, and corrective to, the excesses of progress. It would apply its moderating influence from within medicine, or at least from alongside it. Similarly, the ecological bioethics of Potter or Birch intended to exert moral leadership from within the sciences. But a third kind of bioethics was taking shape at the same

26. See Reich's illuminating discussion in "The 'Wider View,'" 37–41.

time, one that had little sense of partnership with either science or medicine: a feminist bioethics.

There is something rather obvious about feminist interest in ecology. It was expressed elegantly by Ynestra King, who wrote:

> All human beings are natural beings. That may seem like an obvious fact, yet we live in a culture that is founded on the repudiation and domination of nature. This has a special significance for women because, in patriarchal thought, women are believed to be closer to nature than men. . . . The hatred of women and the hatred of nature are intimately connected and mutually reinforcing.[27]

This goes beyond the kind of ecological bioethics described so far in that it calls into question the moral innocence of the players in the world of science, technology, and industry. Potter seemed to believe that once scientists understood the relation between their activity and the values that in another part of their lives they held dear, they would change the way they worked and create the "more and better science" needed to avoid global catastrophe. Birch and Cobb recognized more clearly how deeply entrenched was the anti-life mechanistic model of the world, and granted that a paradigm shift would be necessary if the ecological model was to be embraced and a more humane and sustainable society created. But the essential change was still in the realm of ideas. Change the way you think and you will change the way you live.

The liberal humanist strategy is challenged radically by King's use of the word "hatred." Hatred is not a thought or an idea, but something more primal and far less plastic. A bad idea can be changed into a better idea or even a new idea. But hatred can only be fought, contained, and, if possible, rooted out.[28] Feminist bioethics begins from the realization that the struggle is not primarily over ideas. It is over relationships—social, political, economic, and all of them gendered.[29] So Ynestra King laid down the challenge to the environmental movement: "If male ecological scientists and social ecologists fail to deal with misogyny—the deepest manifestation of nature-hating in their own lives—they are not living the ecological lives or creating the ecological society they claim."[30]

27. Ynestra King, "The Ecology of Feminism and the Feminism of Ecology," in *Healing the Wounds: The Promise of Ecofeminism,* ed. Judith Plant (Philadelphia and Santa Cruz: New Society Publishers, 1989) 18–28, at 18.

28. Or, in terms of the Christian hope, hatred is to be met by that redemptive love that achieves repentance and reconciliation, once and for all in the passion, death, and resurrection of Christ and witnessed to in the nonviolent direct action of Christ's followers.

29. See Ariel Salleh, *Ecofeminism as Politics: Nature, Marx and the Postmodern* (London and New York: Zed Books, 1997) 4–7.

30. King, "The Ecology of Feminism," 24.

Medical bioethics came under a similar critique. Asking why medical ethics was failing to heal medical practice, feminists would observe that the new bioethics normally took the medical institution as a given in ethical discussions, which then focused on particular activities going on within that unexamined, unquestioned structure. Yet even a cursory glance at the institution would indicate that medical practice was dominated by men and reflected men's perspectives and values. It was also clear that women and their dependent children constituted a disproportionately large percentage of the consumers of medical services. Intuitively, this situation was seen to pose risks for women and, indeed, analysis found that these risks were realized in systemic patterns of misdiagnosis, mistreatment, and the unethical use of women seeking health care.[31]

As ecofeminism took shape in the mid-seventies, it included a consistent critique of the institutions of medicine. Rosemary Radford Ruether devoted a chapter of her classic *New Woman New Earth* to the exposure of the sexist ideology that shaped the psychoanalytic revolution.[32] Seeing modern medicine as a whole directed toward women by men, Mary Daly's *Gyn/ecology* used the term "gynaecology" to refer to all the branches of scientific medicine. And when she discussed gynaecology in the narrower sense of "specialized treatment for women," Daly was careful to remind the reader that "the purpose and *intent* of gynecology was/is not healing in a deep sense but violent enforcement of the sexual caste system."[33] In the radical analysis that was being developed within ecofeminism, medical bioethics was to be challenged because it did not address the sickness that was at the root of scientific medicine itself, a "hatred" of the bodies of which it purported to be the savior, and of women's bodies in particular.[34]

31. See, for example, Susan Sherwin, "Feminist and Medical Ethics: Two Different Approaches to Contextual Ethics," in *Feminist Perspectives in Medical Ethics,* ed. Helen Bequaert Holmes and Laura M. Purdy (Bloomington and Indianapolis: Indiana University Press, 1992) 17–31, at 22–23; Barbara Hilkert Andolsen, "Elements of a Feminist Approach to Bioethics," in *Feminist Ethics and the Catholic Tradition: Readings in Moral Theology No. 9,* ed. Charles E. Curran, Margaret A. Farley, and Richard A. McCormick (New York/Mahwah, N.J.: Paulist Press, 1996) 341–382; and Margaret A. Farley, "Feminist Theology and Bioethics," in *Feminist Theological Ethics: A Reader,* ed. Lois K. Daly (Louisville, Ky.: Westminster/John Knox Press, 1994) 192–212.

32. Rosemary Radford Ruether, *New Woman New Earth: Sexist Ideologies and Human Liberation* (San Francisco: Harper & Row, 1975) 137–159.

33. Mary Daly, *Gyn/ecology: The Metaethics of Radical Feminism* (London: The Women's Press, 1989 [1979]) 227.

34. Marti Kheel, "From Healing Herbs to Deadly Drugs: Western Medicine's War Against the Natural World," in *Healing the Wounds,* 96–111. The most public manifestation of the ecofeminist critique of biotechnology was the Feminist International Network of Resistance to Reproductive and Genetic Engineering (FINRRAGE). It is best known for its critique of

Those themes that were common to medical and ecological bioethics were also found in feminist bioethics, only in a more radical and confronting form. Not only was the possibility of "value free" science rejected but the pathological violence of "objective science," against both nature and women, was exposed and struggled with. In this context feminist bioethics challenged the domination of ethical and political discussion by "experts," revaluing the wisdom and knowledge of those who had been the objects of this violence and the most practiced in its resistance. So, too, feminist bioethics tended to be skeptical about the abstract, principalist approach to the new discipline for its failure to deal adequately with the necessary contextuality of moral reflection and the narrative structure of women's moral reasoning. Again, the feminist affirmation of personhood in relationality tended to take sexism as its point of departure, exposing the way patterns of discrimination and oppression systematically distorted the personhood of those involved in the relational web.

Similarly, the focus on embodiment began characteristically with a critique of the mind-body, spirit-flesh, reason-passion, man-nature dualism, which had always associated women with the subordinate position. Against this, feminism called into question the pattern of dualism itself, emphasizing the necessary interconnection between the elements in each binary pair and, at the same time, honoring and celebrating the life of the body in all its dimensions. Finally, the emphasis of bioethics on the connectedness of human beings with other species and entities was radicalized in feminism as the solidarity of women with the earth and all its creatures, in resistance to a common assailant: "misogyny—the deepest manifestation of nature-hating in their own lives."

It hardly needs to be said that to the extent that feminist bioethics employed theological points of reference, it never approached ethics as any kind of *applied* theology. Indeed, its theological impulse was not just revisionist but radical and constructivist. It included at once a radical denunciation of those aspects of the Christian tradition that were complicit in the structure of misogyny and a bold annunciation of alternative, healing apprehensions of God. As is well known, there is no single, uniform feminist theology. And the theologians associated with feminist bioethics

---

reproductive technologies. See, for example, Robyn Rowland, *Living Laboratories: Women and Reproductive Technologies* (Sydney: Pan Macmillan, 1992); Janice G Raymond, *Women as Wombs: Reproductive Technologies and the Battle Over Women's Bodies* (New York: HarperCollins, 1993); and, for a critique, Max Charlesworth, "Whose Body? Feminist Views of Reproductive Technology," in *Troubled Bodies: Critical Perspectives on Postmodernism, Medical Ethics, and the Body,* ed. Paul A. Komesaroff (Melbourne: Melbourne University Press, 1995) 125–141.

may be located on a continuum, from the radicalism of Rosemary Radford Ruether, to the constructivism of Sallie McFague or Mary Grey, to the renewed classicism of Elizabeth Johnson or Celia Deane-Drummond.[35] And it is no coincidence that there is a chronological dimension to this continuum too. The agenda of feminist theology has shifted as the ecclesial and social context has changed and, especially, as the central claims of feminism have won broad acceptance in the theological community, an acceptance reflected in the contributions to this Ecotheology Project.

Theology, Ecology, and Bioethics

For much of the eighties and early nineties, ecological bioethics and medical bioethics worked without reference to each other. Ecological bioethics tended to be called "environmental ethics," while it was generally accepted that "bioethics" referred quite narrowly to the medical sphere. Moreover, so much was happening in each area that the lack of communication between them hardly mattered. Ecological bioethics was preoccupied with newly recognized phenomena such as the greenhouse effect, requiring new accounts of the obligations we have to our neighbors, regardless of political and even geographical frontiers. In medical bioethics new developments in, for example, reproductive technology required a reconsideration of the nature of human life, bodily integrity, and the idea of family.

But, increasingly, there are developments in science and technology that bring the two streams together again. Some of these developments are fairly obvious and invite a simple comparison of insights. For example, there are clear links between environmental degradation and certain human illnesses, such as the correlation between deforestation and the spread of malaria or between ozone depletion and the incidence of melanoma and other skin cancers.[36]

But other developments in science and technology make the distinction between ecological and medical bioethics counterproductive and questionable in itself. For example, the same genetic sciences are used in, say, the genetic modification of non-human species in agriculture and the

35. See, for example, Rosemary Radford Ruether, *New Woman New Earth;* Sallie McFague, *Models of God: Theology for an Ecological, Nuclear Age* (London: SCM Press, 1987); Mary Grey, *Redeeming the Dream* (London: SPCK, 1989); Elizabeth A. Johnson, *She Who Is: The Mystery of God in Feminist Theological Discourse* (New York: Crossroad, 1992); Celia Deane-Drummond, *Theology and Biotechnology: Implications for a New Science* (London: Geoffrey Chapman, 1997).

36. See A. J. McMichael, *Planetary Overload: Global Environmental Change and the Health of the Human Species* (Cambridge: Cambridge University Press, 1993).

diagnosis of genetic diseases in human embryos and fetuses. Any bioethical discussion of the values associated with the genome *as such* will have to involve both medical and ecological perspectives from the outset. Or again, in the emerging field of bioinformatics, computer science and genetic science are being brought together to achieve results that were the stuff of science fiction only a decade ago. For example, DNA has been placed on silicone chips to create the "DNA chip," which is expected to be able to read all the information in the genomes of living organisms. Its diagnostic potential is enormous, and so is the complex of moral quandaries it creates for patients, doctors, families, the general public, and legislators. And already lessons being learned in this technological advance are being applied in the development of the "molecular computer," in which silicon is to be replaced entirely by DNA strands to create a "thinking machine" far more powerful than today's supercomputers. Distinctions are blurring and disappearing between the human and non-human, the organic and inorganic, the natural and artificial, and even between reality and its computer-generated "virtual" equivalent.[37]

The moral dimensions of these head-spinning developments are made even more complex by the realization that research and development are largely driven by the pursuit of profit rather than by a concern for the common good. As Christine Burke has shown in Chapter 2 of this book, making ethical sense of our situation requires an analysis of economic globalization and the eclipse of nations and democracy. The most pressing bioethical issues are clustered around the activities of multinational pharmaceutical companies and agribusiness.

Today, every bit as much as at the time of its emergence in the early seventies, it is necessary that bioethics be oriented toward a field much broader than medicine as it applies itself to its role as an agent of life and healing. And just as it did when bioethics was first taking shape, theology has a useful contribution to make to that service of life and healing. If anything, theology is now better suited to a collaboration with bioethics. It has certainly moved beyond seeing bioethics as a kind of applied theology to recognize and welcome the way bioethics facilitates the review and revision of theology's methods and principles in conversation with the life sciences. So, too, after a generation of sustained critique by feminist and other radical perspectives, it has learned to be more open, expecting fresh insights from new participants. In this way theology has attained a new fluency in the language of spirituality, pastoral analysis, and ethics. The contributions to the present Ecotheology Project include many examples

---

37. Jeremy Rifkin, *The Biotech Century: Harnessing the Gene and Remaking the World* (New York: Penguin Putnam, 1998) 226–237 and passim.

of theology's readiness to participate in the different kind of bioethics demanded by our changed global circumstances.

To begin with, ecologically oriented theology contests the isolation of ethical reflection on the life of the body from ethical reflection on the life of the earth. So, for example, in his Christological essay in Chapter 4 of this volume, Duncan Reid has worked toward the conclusion that "the concerns of all flesh will take logical priority to the concerns of the human, as they do in the Creed, even though the concerns of humanity may be allowed, as they do in the Creed, to have the final say." The ethical import of his doctrinal analysis is that bioethical reflection must locate human interests in the context of the interests of other species and entities. Reflection on the healing of human beings necessarily involves reflection on the healing of the earth and all its creatures. So, too, Denis Edwards's pneumatological essay in Chapter 3 of this book argues that all God's creatures, and not just the human species, are the object of the Holy Spirit's distinctive work of "empowering and communion-making." So while different creatures may have different immediate goods, they share as one complex community of creatures the ultimate good of communion within the relations of the Trinity. The soteriological insights that form the "central panel" in Lucy Larkin's essay in Chapter 8 name this ultimate good as "healing":

> Salvation is God's healing work . . . . God's intimate and integral relationship to me as a living being means that salvation is *also* brought about when I reground myself in the healing relationality. That is, to acknowledge my part in a continuous process of creative transformations toward the wellbeing of all forms of life is also to enrich the life of God. Healing, ecological or otherwise, comes about through this series of transformations, which are a co-creation of Creator and creature.

Thus, from the theological perspective being explored in this Project, a bioethics which treats the life of the body in isolation from the life of the earth and which quarantines the consideration of human interests from those of non-human creatures is highly problematic, indeed incoherent. It is no longer just a quirk of history—there is something wrong with the *idea* of a bioethics that is concerned about human health and healing but not about the health and healing of the earth.

The theological anthropology that has emerged in our Ecotheology Project points in the same direction. Patricia Fox, for example, in Chapter 5, has argued for a Trinitarian anthropology according to which "to be a person is to be in dynamic mutual relation with other persons and created entities." So, too, James McEvoy's use of the relational philosophy of Charles Taylor in Chapter 11 steers us toward the point of saying that interpersonal relationships are *constitutive* of the human being.

Medical bioethics provides a way for us to give expression to this perspective in the area of health care, and, indeed, some of the most important theological contributions to medical bioethics have been in ecclesiology. For example, in his classic essay "Why Medicine Needs the Church," Stanley Hauerwas argues that "given the particular demands put on those who care for the ill, something very much like a church is necessary to sustain that care."[38] By "a church" he does not mean primarily an institution or hierarchy—he means a community of Christians. The "particular demands" to which he refers are the demands of caring for the sick by being present to them in their pain, regardless of the likelihood of being able to "cure" the person and for as long as that person suffers. The Church at least claims to be "a people who have so learned to embody such a presence in their lives that it has become the marrow of their habits."[39] Hauerwas's view is not that medicine needs the Church as an ethical teacher, but that those who care for the sick need the Church (or "something very much like a church") as a community that can model and resource the kind of presence to suffering that is the essence of the healing vocations.

But medical bioethics also raises problems for our theological approach. To emphasize relationality as the basis of personhood to the extent that we do is to risk being misunderstood in a quite specific way. In the popular imagination, and in some less careful critical studies, relationality is irreducibly psychological. In that case the personhood of certain human organisms comes into dispute just because they are not capable of being in relationship, for example, the comatose patient, the anencephalic newborn, the surplus human embryo. It was out of concern for examples such as these that Ramsey tended to emphasize embodiment even more than relationality in his discussions of personhood, to the extent that he leaned toward the now rather old-fashioned idea of a bodily *imago Dei*.

In our Project, Lorna Hallahan's theological engagement with the "dungheap" has required that we do not lose sight of the way that persons are known bodily and, indeed, even in bodily failure. She has called us to reflect on "our impulse to cure," inviting a reconsideration of the historical fact that while Christianity has no vested interest in the *cure* of the sick, it is ideologically committed to the *care* of the sick.[40] "Our hope," she says, referring again to that ultimate good of healing in communion, "lies in the offer of embrace of all those creatures and places trapped in unloveliness."

38. Stanley Hauerwas, *Suffering Presence: Theological Reflections on Medicine, the Mentally Handicapped, and the Church* (Edinburgh: T&T Clark, 1986) 63–83, at 75.

39. Ibid., 80.

40. See Gary B. Ferngren, "Medicine and Compassion in Early Christianity," *Theological Digest* 46 (Winter 1999) 315–326.

It is not unexpected that there would be a congruence between the emphases of ecological bioethics and elements of our Ecotheology Project. That is particularly so in the case of our conviction that not only human beings but all creatures are caught up in the life of the triune God through the incarnation, death, and resurrection of Christ and the Holy Spirit's ongoing work of empowerment and creating communion. The human chauvinism of received Christianity is under sustained critique here. But this point of congruence is not entirely comfortable for us. While we are ready to affirm the intrinsic value of non-human creatures, we have not attempted to enter into a consideration of questions such as vegetarianism, the use of animals in research, the use of genetically modified animals in medicine, or genetic engineering in agriculture. However, ecological bioethics insists that such questions must move to the front of our minds so that we become more specific about precisely what it means for non-human species and entities that they are God's beloved creatures.

It is only to be expected, however, that there will be open questions and points of discomfort in the engagement of theology with bioethics. For bioethics is not only a language in which theology can critique life-denying tendencies in technologically advanced societies; bioethics is equally a language in which theology hears a critique of its own life-denying tendencies.

Index

universe, xv, xvi, 40, 47, 48, 52, 55,
   65, 76, 78, 95, 96, 99, 126, 127,
   129, 130, 133, 135, 136, 138, 139,
   141–143, 149, 150, 156, 162, 163,
   166, 167, 170, 172, 183, 184, 209
unloveliness, 105–123, 229

Verhey, A., 213n
Visick, V., 160n
Volf, M., xv, 106, 114, 116–120, 121n,
   145n, 152n
Voltaire, 199

Waldstein, M., 137n
Walker, A., 121
Wallace, M., 71n
Warren, K., 155n
Weber, M., 204
Weil, S., 108, 120n
Welch, S., 110, 111, 118, 119
Wells, H., 161n
Westcott, B., 218n
western
   advances, 27

culture, xiii, 201
theology, 58
tradition, 58
Wheatley, M., 38
White, L., 12
Whitehead, A., 147n, 184, 185
wilderness, 8
Wilson, R., 98n
Wilson-Kastner, P., 132
Winterson, J., 114n, 115
Wolseley, J., 17
Woolfolk, R., 197n
Word, xiv, 3, 57, 59, 60–62, 75–80, 83,
   134, 136, 139, 164, 195, 209
world, 178, 182, 185, 190, 191, 195,
   197, 201, 223
   natural, 1, 5, 6, 8, 9
   trade, 23
Worster, D., 8n
Worthing, M., 195n

Zizioulas, J., xv, 47, 74n, 81n, 87–93,
   95, 99–103, 130–133, 138, 142n,
   146n, 148n